Materials in Construction

An Introduction

:002

:

Materials in Construction

An Introduction

Third edition

G.D. Taylor
PhD, CertEd, MCIOB

LONGMAN

Pearson Education Limited
Edinburgh Gate, Harlow
Essex CM20 2JE, England
and Associated Companies throughout the world

© Longman Group UK Ltd 1985, 1994
© Pearson Education Limited 2000

First published 1985
Second edition 1994
Third edition 2000

British Library Cataloguing in Publication Data
A catalogue entry for this title is available from the British Library

ISBN 0-582-36889-8

Set by 35 in Times 10/12 and Frutiger
Printed in Singapore

Contents

Acknowledgements   ix

1   Introduction   1
1.1   Stages in the life of a building   1
1.2   Performance criteria   2
1.3   Materials properties: specifications, standards and regulations   3
1.4   Quality, quality control and quality management   6
1.5   Materials performance and its measurement   8
      Self-assessment questions   20
      Standards   21

2   Structure and performance of materials   22
2.1   Introduction   22
2.2   The atom   22
2.3   Atomic bonding   28
2.4   Bond strengths and strengths in practice   32
2.5   More recent materials   34
      Self-assessment questions   40

3   Concrete   41
3.1   Introduction   41
3.2   Cements   42
3.3   Structure of hydrated cement   50
3.4   High-alumina cement   50
3.5   Aggregates for concrete   51
3.6   Water for concrete   60
3.7   Principles of proportioning concrete mixes   60
3.8   Prescriptive concrete mixes   67
3.9   Designed concrete mixes   68
3.10 Department of the Environment method of mix design (1988 edition) 69
3.11 Curing of concrete   80
3.12 Statistical interpretation of cube results   81
3.13 Shrinkage and moisture movement of concrete   83
3.14 Durability of concrete   85
3.15 Corrosion of steel reinforcement   88

3.16	Admixtures (BS 5075)	90
3.17	Lightweight concretes	93
3.18	High-strength concrete	95
3.19	Production of high-quality surface finishes	96
3.20	Recycling concrete	98
	Experiments	100
	Self-assessment questions	110
	Reference	113
	Standards	113

4	**Masonry**	115
4.1	Introduction	115
4.2	Bricks	116
4.3	Building blocks	129
4.4	Natural stonemasonry	132
4.5	Masonry mortars	134
4.6	Strength of masonry	138
4.7	Renderings	139
4.8	Dampness in masonry	141
4.9	Recycling masonry	152
	Experiments	153
	Self-assessment questions	160
	References	161
	Standards	161

5	**Glass in construction**	163
5.1	Introduction	163
5.2	Manufacture	164
5.3	Strength of glass	165
5.4	Safety with glass (BS 6206)	166
5.5	Thermal performance	169
5.6	Decorative use of glass	172
	Self-assessment questions	173
	Standards	173

6	**Plasters**	174
6.1	Introduction	174
6.2	Plaster types	175
6.3	Classes of gypsum plaster (BS 1191)	175
6.4	Lightweight aggregates	177
6.5	Lime	177
6.6	Plaster systems	177
6.7	Plasterboards	179
6.8	Plaster for damp applications	181
6.9	Common defects in plastering	181

	Self-assessment questions	183
	Reference	183
	Standard	183

7	**Metals**	184
7.1	Introduction	184
7.2	Metal structure	185
7.3	Steel	190
7.4	Steel types used in construction	195
7.5	General applications of metals in construction	201
7.6	Corrosion of metals	205
7.7	Recycling metals	212
	Experiments	213
	Self-assessment questions	215
	Standards	217

8	**Timber**	218
8.1	Introduction	218
8.2	Structure of wood	219
8.3	The effects of moisture in timber	223
8.4	Conversion of timber	226
8.5	Factors affecting the distortion of timber prior to use	227
8.6	Durability of timber	228
8.7	Preservative treatments for timber (BS 1282)	231
8.8	Current preservation requirements for new building work	234
8.9	Appropriate preservation levels	238
8.10	Fire resistance of timber	239
8.11	Timber for structural uses	242
8.12	Timber products	255
8.13	Environmental considerations in the use of timber	260
	Experiments	263
	Self-assessment questions	271
	References	273
	Standards	273

9	**Plastics**	275
9.1	Introduction	275
9.2	Importance of carbon	276
9.3	Polymerisation	279
9.4	Classification of plastics	280
9.5	Some common thermoplastics	284
9.6	Elastomers	287
9.7	Some common thermosets	289
9.8	Plastics in fire	291
9.9	Application of polymers in the construction industry	293

9.10   Recycling plastics   294
       Self-assessment questions   297

10   Paints   298
10.1   Introduction   298
10.2   Painting systems   299
10.3   Some common types of paint   300
10.4   Painting specific materials   302
       Self-assessment questions   306
       Standards   306

11   Bituminous materials   307
11.1   Introduction   307
11.2   Bitumen (BS 3690)   308
11.3   Road tar (BS 76)   311
11.4   Principles of design of road pavements   312
11.5   Asphalts and macadams   314
11.6   Surface dressings   317
11.7   Defects in bituminous pavements or surfacings   318
11.8   Bituminous products for roofing   319
       Experiments   322
       Self-assessment questions   327
       References   328
       Standards   328

Answers to numerical self-assessment questions   329
Additional reading   331
Index   332

Acknowledgements

The author gratefully acknowledges the support of colleagues, friends and family for their help during the research and production of this book.

The publisher wishes to thank the Building Research Establishment for permission to reproduce our Figs 3.12, 3.13, 3.14, 3.15, 3.16, 3.17 & 8.16, all adapted from BRE materials.

Extracts from British Standards are reproduced with the permission of BSI under licence number PD 1999 1102. Complete copies of the individual standards can be obtained by post from BSI Customer Services, 389 Chiswick High Road, London W4 4AL. [Our Figs 1.8 & 5.3 (both adapted from BS 6206: 1981) and our Tables 3.2 (adapted from BS 12: 1996), 3.7 (adapted from BS 5328 Part 2: 1997), 4.4 (adapted from BS 5628 Part 2), 8.3 (adapted from BS EN 335-1), 8.4 (adapted from BS EN 460: 1994), 8.5 (adapted from BS 4978: 1996), 8.6 & 8.7 (both adapted from BS 5268 Part 2: 1996), 8.11 (adapted from BS EN 622-2/3/4/5, BS EN 312-1/4/5/6 & BS EN 300).]

1 Introduction

Chapter summary

- Stages in the life of a building.
- Performance criteria, standards and quality in relation to materials in construction.
- Important performance characteristics of materials, together with means of assessing them.
- Important environmental performance issues.

1.1 Stages in the life of a building

The successful operation of materials in buildings requires an understanding of their characteristics as they affect the building at all stages of its lifetime as follows:

- **Design**
  Materials appropriate to the building design, function and environment must be selected; indeed, traditionally, consideration of the materials to be used was a coherent part of the design process. Today there is more choice but designers must be aware of the limitations as well as the opportunities associated with individual materials, and how they interact with each other.

- **Construction**
  Builders are legally obliged to 'build well' and this includes a responsibility to use specified materials in the correct manner and to identify potential defects.

- **Maintenance**
  Although some materials are virtually maintenance-free, the majority require some form of care during their lifetime. Effective maintenance depends upon knowledge of how materials react with their environment over the planned lifetime. It may in the long term be much more expensive to maintain certain materials if their initial quality is poor – for example, protective coatings to timber and metals.

- **Repair**
  A wide choice of repair systems is now available for many materials. Informed and efficient use of such systems hinges upon an understanding of decay processes, how to arrest decay, and how the repair materials interact with the original materials.

- **Demolition/recycling**
  These processes are increasingly being considered as a part of the overall selection procedure. The safety/environmental aspects of alternative materials are, in particular, now being included in assessments of overall suitability.

1.2 Performance criteria

Modern buildings are usually one-off structures of great complexity. They comprise many components, some pre-manufactured and assembled on site, while others, such as concrete, may be manufactured 'in situ'. In each case, satisfactory operation of the building as a whole depends on the performance of the materials from which its components are made and on the interrelationships between them.

Before assessing the suitability of any one material for a given situation, the **performance** requirements for that situation must be identified. Such requirements might include:

- **Structural safety**
  The ability to withstand stresses resulting from gravity, wind, thermal or moisture movement, or other sources.
- **Fire**
  The material must behave acceptably in resisting fire spread, release of dangerous substances in fire and retaining satisfactory structural stability.
- **Durability**
  The material should fulfil the above performance criteria as required for the planned lifetime of the building.
- **Health/safety**
  There should be no risk to health due to chemical or physical effects of the material, both during and after construction.

There are a number of other important performance requirements; for example:

- Comfort
- Resistance to weathering
- Serviceability
- Appearance

These will be referred to as appropriate when individual materials are described.

In addition to the above performance requirements of materials within the finished building, the following might also have to be considered:

- Availability/cost.
- Ease with which material can be incorporated into the building (buildability).
- Environmental aspects – for example, energy demand of the material during manufacture and ability to conserve energy in use.

1.3 Materials properties: specifications, standards and regulations

Most new buildings are produced with the aid of some written specification, which may often take the form of detailed annotations to drawings, a bill of quantities, or separate documents which are part of the contract. Some form of written specification is highly desirable because it communicates to the construction team exactly what is required and provides the client with a basis on which unsatisfactory performance may be identified and remedied.

It is important to appreciate the background to the production of specifications, standards and, ultimately, regulations for a particular item or context. In order to be enforceable there is a need for each stage of the process to be **precise** and **repeatable**. The developmental stages that lie behind any specific standards or regulations are shown in Fig. 1.1.

Fig. 1.1 *Developmental stages in the production of standards and specifications*

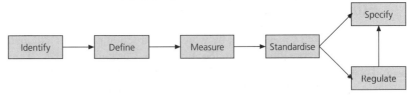

To give an illustration of this, the resistance to fire spread through a walling component might be required. This is the **identification** stage, which might be **defined** in terms of how long (time in minutes) fire would take to penetrate the wall. This would be **measured** by carefully designed experimental tests, but if results are to have widespread applicability, all relevant aspects of the test procedure should be **standardised**. Standards may be confined to the **method** of test, but very often they also include **acceptable levels of performance** under the test.

Finally, a client or designer may **specify** performance levels by reference to the standard or, especially where health and safety are involved, a local authority or government may **regulate** use by requiring a given level of performance under the standard.

Most standards represent basic performance levels. When higher levels are required they may have to be drafted carefully for a given specification. Standards measure performance in a carefully defined reproducible manner, as explained above. They are subject to change as understanding of materials properties increases, experimental techniques improve and performance requirements evolve.

It is important to appreciate that manufacturers may claim compliance with a standard without independent tests being carried out. The best evidence of conformity is obtained when independent tests are carried out by some qualified testing authority. This is the basis of quality marks such as the 'Kitemark' (see below).

British Standards

These standards were produced at the beginning of the twentieth century and now cover a vast range of materials and components, including most of those used in the construction industry. They are produced by committees representing manufacturers, researchers, users and government organisations. With the advent of Eurostandards, British Standards now operate only on a care and maintenance basis – they will eventually be replaced by their European counterparts which have validity in Europe as a whole. Because this process is gradual, many British Standards still exist; indeed new versions may be produced in the transition period (for example, the latest edition of BS 4449 for steel reinforcement was published in 1997 during the preparation of EN 10080, the European equivalent), but such standards will only be valid until the European equivalent is operational. However, the final versions of European Standards, on adoption by member states of the EU in the appropriate language, again become 'national standards'. For example, EN 10080 will be BS EN 10080 and will be available from the British Standards Institution.

European codes and standards

These provide a mechanism by which construction products and materials can be traded and used freely throughout the European Common Market. They are produced by Technical Committees of the European Committee for Standardisation (CEN), which are of international composition reflecting the viewpoints of member states. Although December 1992 was considered to be the implementation date for these standards, their introduction and validation are gradual processes and at present many standards are incomplete. As with British Standards, there is a need for recognised testing and validating authorities in each member state, and this infrastructure inevitably takes time to introduce.

At present there are nine Eurocodes:

1. General and actions
2. Concrete
3. Steel
4. Composite (concrete/steel)
5. Timber
6. Masonry
7. Foundations
8. Seismic design
9. Aluminium

Each Eurocode can be regarded as a code of practice, although there is normally nothing to prevent individual countries continuing with their national codes for the time being. Eurocodes are being supported by European Standards (ENs) whose use is enforced by **directives**. The **Construction Products Directive (CPD)** was given legal force in 1991. Associated standards are based on six 'essential requirements' relating to health, safety and energy conservation, as defined in the CPD. They are:

- Stability
- Safety in fire
- Safety in use
- Health
- Noise
- Heat retention

Fig. 1.2 *CE mark as evidence of conformity of a product to an EC directive.*

As with British Standards, European Standards must be developed with great care, representing the viewpoints of various 'interested' parties. They are therefore produced first as prestandards, or ENVs, which have no legal status; these being followed by the full standard after a period of about three years. Any product which complies with the essential requirements given in the standard is eligible to carry a 'CE' mark (Fig. 1.2).

It should be noted that the mark corresponds to **essential requirements** of the standard as defined by the CPD, rather than total requirements. Hence the existence of a CE mark does not necessarily indicate **full** compliance with a given standard. The use of CE marks becomes mandatory once product directives are issued by the European Union. Any standard will require a level of *attestation* which demonstrates that it has been adequately assessed. For non-critical aspects of behaviour, a **manufacturer's declaration** will suffice, while for critical aspects of performance – for example, involving safety – **third party attestation** will be necessary. For example, in the cement standard EN 197–1, third party attestation is essential because the destination of cement is unknown – it could end up in a critical loadbearing situation. CE marking will normally be used in relation to a product which is produced to a quality management system (QMS) – see below. The importance of European standards can be judged from the Building Regulations (1991), Regulation 7 stating that:

> An EC marked material can only be rejected by the Building Control Authority or Approved Inspector on the basis that its performance is not in accordance with its technical specification. The onus of proof in such cases is on the Building Control Authority or Approved Inspector, who must notify the Trading Standards Officer.

Other directives have been issued for various product types – for example, electrical equipment and appliances burning gaseous fuels.

ISO standards

These are standards produced by the International Standards Organisation and they are intended to have world-wide validity. They may be adopted by individual countries as national standards; for example, ISO 9000: *Quality systems*, has been adopted by the BSI and European Union and has the full title BS ISO 9000 (see below).

Agrement certificates

These are produced by the British Board of Agrement (BBA) and generally relate to new products or techniques not covered by a British Standard. Being widely accepted by local authorities, they represent a means of obtaining relatively rapid approval and hence are an encouragement to innovation, though certificates are valid for a limited time period prior to preparation of British or other standards. The BBA operates to QMS principles (see below) but focuses upon the specification, manufacture and performance of the product and its suitability for use in buildings. Production is monitored by inspection, normally on a twice yearly basis throughout the period of validity of a certificate. A European Union of Agrement (UEAtc) operates on a similar basis in other European countries.

1.4 Quality, quality control and quality management

Quality

> Quality can be simply defined as 'fitness for purpose'.

In practice it is necessary to interpret the term precisely in each situation and then to decide how 'quality' requirements should be implemented, bearing in mind that there will always be a cost implication as target quality levels rise.

The first stage in considering a specification is to arrive at an appropriate/realistic target for quality. Some factors to be considered when arriving at such a target level for a specific item are:

1. What are the possible failure modes?
2. What are the consequences of failure in safety terms?
3. How easy is it to inspect/maintain the item?
4. How easy/costly would it be to replace the item if it failed?

Quality control

Quality control is the practical procedure which assists in the production of a quality product. In the contexts of items produced in a factory or on site, there are numerous opportunities for defects to arise by machine or by human error. Quality control attempts to identify these possible events and implements procedures to avoid them if possible and to detect and rectify them should they occur.

In human terms, the mere fact that a process is being monitored is likely to assist in avoiding mistakes, especially if penalties are imposed for such errors. It is important to appreciate that, in this technological age, personnel rarely have a full understanding of the manufacturing process in which they are involved, or of all the possible consequences of any error. In this situation the importance of adequate supervision will be evident. In any one context the level of supervision to be invoked must be decided upon by consideration of the performance and reliability required of the item being produced, together with the overall cost of a failure including damage to reputation, compared with the cost of providing a given level of quality control.

Quality management system

Quality management involves the operation of a comprehensive system of quality control, including employment of a quality manager to oversee the maintenance of quality standards and keeping of systematic written records of every part of a design, production, or other process.

The best way of operating quality management is by means of a recognised quality management system (QMS). The first widespread system to be operated was BS 5750 from 1979. BS 5750 was adopted by ISO as ISO 9000 in 1987 and the harmonised version, produced in 1994, was published as **BS ISO 9000**.

Although companies are not required to register a QMS with a third party, such as an independent certification body, there are obvious advantages in doing so – for example, public recognition. The certification bodies themselves must be accredited and this is organised through the National Accreditation Council for Certification Bodies (NACCB). The British Standards Institution is a widely used QMS accreditation body and currently has the most QMS registrations, although other important accreditation bodies in the construction area include Lloyd's Register of Quality Assurance, the Quality Scheme for Ready Mixed Concrete and the UK Certifying Authority for Reinforcement Steels. Companies become 'registered' under the standard if they satisfy the necessary criteria.

QMS is more readily applied in factory situations where operations are more mechanised and repetitive than in site situations, although designers, building and other contractors, and those providing services are rapidly moving towards quality systems of this type, having regard to their benefits, both in lower failure rates and marketing advantages.

BS kitemark

The British Standards Institution awards the kitemark (Fig. 1.3) to products produced and assessed to a quality system operated to BS EN ISO 9000. The kitemark is most commonly awarded against British standards, but it can be obtained against any national, European or ISO standards for the product. It has widespread recognition and is synonymous with high standards of production and quality control.

Fig. 1.3 _The kitemark as evidence of standard conformity by an independent test authority._

1.5 Materials performance and its measurement

The main considerations under which performance might be assessed were mentioned at the beginning of this chapter. The following is an amplification of some of these areas in order that the critical parameters in any one material might be identified and controlled:

- Stability (structural)
- Safety in fire
- Durability
- Safety in use and health
- Environmental issues

Stability (structural)

There are many aspects of structural performance; the main ones being as follows:

Strength

Strength may be defined as the ability to resist failure or excessive plastic deformation under stress.

There are several types of stress (Fig. 1.4). In general, the strength type selected should be as close as possible to that experienced in the structure, but even then, care is necessary with all types of strength measurement.

Fig. 1.4 _Basic stress configurations._

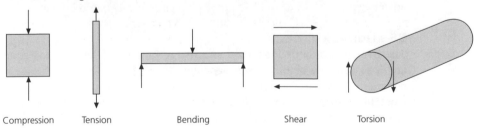

Compression   Tension   Bending   Shear   Torsion

Fig. 1.5 *Effect of specimen shape on compression test result.*

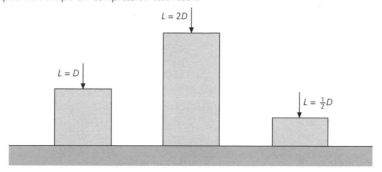

Fig. 1.6 *Shear stress induced by applied compressive stress.*

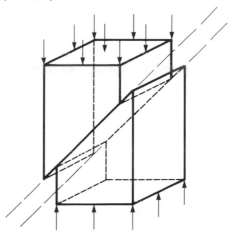

Compression testing needs special care because:

- Slender specimens may tend to buckle, and this is really a failure in bending.
- The result depends upon the frictional restraining force between the platen of the testing machine and the specimen which, in turn, depends on the design of the machine.
- Compression test results depend on the shape of the sample. The situation is represented in Fig. 1.5, which shows that highest strength for three compressive specimens of equal cross-sectional area will be the sample with $L = \frac{1}{2}D$ since the platens of the testing machine 'reinforce' the material as it tends to spread sideways under load. The specimen with $L = 2D$ will record the lowest 'stress'.

It may well be argued that a pure compression test should not cause failure since the forces are assisting atomic bonds and should therefore tend to hold the materials together. While this is true, **uniaxial** compression (compression in one direction only) also induces shear forces (Fig. 1.6) and tensile stresses as the material swells in a lateral direction.

Tension testing can be difficult, especially with materials such as ceramics because gripping specimens of uniform section produces stresses within the grips which tend to cause premature fracture there (Fig. 1.7). Such problems can be overcome by, for example, waisting the specimen, but this adds to the cost of the test.

Fig. 1.7 *Additional stresses acting on a specimen when tensile stress is applied via friction grips. (The diagram is schematic only; the friction grips would operate, for example, by pivots and a scissors actions.)*

Bending produces compression, tension and shear stresses within the material, the balance between these depending upon the shape of the sample. These stresses can be calculated but the calculations presuppose simple bending theory and it is known that such theory cannot be applied accurately to many materials (such as timber and concrete).

Torsion, or twisting, occurs quite commonly when loads are applied eccentrically. For example, if a beam supports a further beam at right angles to it at mid-span, the carrier beam will be under a torsional stress. The calculation of torsional stresses and resistance of materials to torsion can be quite complex.

All stresses should be measured in newtons per square mm (N/mm^2). Note that fracture is not necessary for a strength 'failure'. For example, steel for practical purposes, has failed when it 'yields', though yielding does not initially damage the metal. There are, on the other hand, many materials – ceramics, for example – where sudden fracture is the first indication that strength has been exceeded.

Stiffness

Stiffness is the ability of a material to resist elastic deformation under load.

Elastic deformation is deformation which is recovered when the load is removed. High deformations, even if elastic, may cause problems – for example, unsightly appearance or failure of plaster coatings if attached to a substrate which has large deflections. It is normal practice, therefore, to check deflections as part of the structural design. Stiffness is normally measured by:

$$\text{elastic modulus } (E) = \frac{\text{applied stress}}{\text{strain caused by that stress}}$$

Since strain has no units (= fractional change of length), E values have the units of stress. Measured strains are usually small, typically less than 0.001, leading to large values of E. They are, therefore, usually measured in kN/mm^2.

$$1 \text{ kN/mm}^2 = 1000 \text{ N/mm}^2$$

Toughness

Toughness is the ability of a material to absorb energy by impact or sudden blow.

Strong materials are not always tough – for example, cast iron. Relatively weak materials can have high toughness – for example, leather. Figure 1.8 illustrates a typical impact test such as that used to measure the behaviour of toughened glass.

Fig. 1.8 *Impact tester suitable for testing toughness of glass. (Adapted from BS 6206: 1981)*

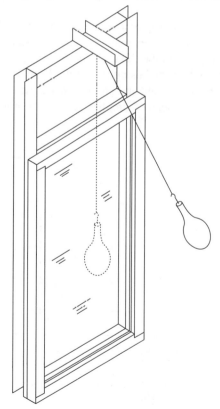

Hardness

Hardness is resistance to indentation under stress.

Hardness is relevant to floor and wall surfaces (Fig. 1.9) and depends on a combination of strength and stiffness properties.

Fig. 1.9 *Hardness testing.*

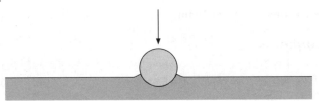

Creep

Creep is the effect of long-term stress, leading to additional distortion or failure.

Creep does not necessarily produce a safety risk though marked deflections in, say, beams can appear alarming. In extreme cases, creep may lead to ultimate failure. Materials subject to creep are timber, clay, lead, concrete, thermoplastics and, to a small extent, glass. Creep can be tested using rigs of the form shown in Fig. 1.10. Simple mechanical systems are preferred and the strain can be measured by a demountable dial gauge. Alternatively, deflections of simply supported beams such as timber could be measured.

Fig. 1.10 *Creep testing.*

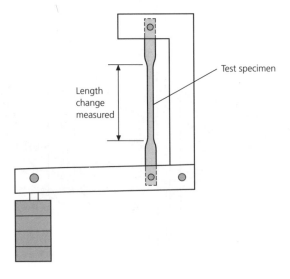

Length change measured

Test specimen

Fatigue

Fatigue is the effect of load reversals such as vibrations which lead to failure at relatively low stresses.

All materials are subject to fatigue effects and, in some situations – for example, roads or floors subject to heavy moving loads, or machine frames – fatigue may be the critical factor in design. The fatigue life of components is measured in cycles since each cycle adds to the level of damage. For metals, fatigue lives of several million cycles may be required in situations at risk.

Comparative properties of construction materials

See Table 1.1. The figures presented in this table are simply a guide, and should only be used for comparative purposes.

Table 1.1 *Common construction materials: structural properties*

In order of increasing density

Material	Density (kg/m³)	Tensile strength (N/mm²)	Elastic modulus (kN/mm²)	Toughness	Hardness	Creep resistance
Softwood (pine) || to grain	500	90	9	High	Low	Low
Hardwood (oak) || to grain	700	140	10	High	Moderate	Moderate
Elastomer (butyl rubber)	900	10	0.1	High	Very low	Low
Thermo-plastic (rigid PVC)	1100	50	2.5	Moderate	Moderate	Moderate
Clay facing brick (stock)	1900	3	20	Low	High	High
Concrete (normal)	2300	3	30	Moderate	High	Moderate
Glass (glazing)	2600	100	70	Low	High	Moderate
Glass (fibre reinforcement)	2600	3000	70	Moderate	High	High
Steel (mild)	7800	300	210	High	Very high	Very high
Steel (prestressing)	7800	2000	210	High	Very high	Very high

Safety in fire

Combustion is a process involving chemical reaction of a fuel (usually organic material containing carbon) with oxygen in the presence of heat.

To initiate the process, heat or a source of ignition is essential, though, once started, many combustion processes are self-sustaining because heat is a by-product. The situation is illustrated by the triangle of Fig. 1.11.

Fig. 1.11 *The fire triangle: if any one ingredient is missing the fire cannot start.*

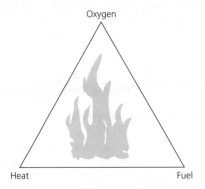

To prevent or extinguish a fire, it is only necessary to remove one of the three essential ingredients; for example, fire risk is greatly reduced as the quantity of combustible material in an enclosure ('Fire Load' – Building Regulations) is reduced. Some extinguishers work by surrounding the fire with a gas such as a 'halon', containing the halogens chlorine, fluorine or bromine which interfere with the combustion process/prevent oxygen access. Water extinguishes fire by removal of heat, greatly assisted by vaporisation of water which is a highly endothermic reaction, consuming very large quantities of heat.

The calorific value of a fuel measures the heat output on total combustion and does not vary greatly within the range of organic materials. For example, wood, coal, paper, petrol are all in the range 15–50 MJ/kg. It should be appreciated that a 'flame' is burning vapour and for it to occur the burning material must vaporise. The risk in practice varies greatly, being greatest with liquid or gaseous fuels. Liquid fuels are volatile – they vaporise very easily. Together with gaseous fuels they therefore mix easily with oxygen. The risk with bulk solid materials such as large timber sections is lower since oxygen access is limited and they vaporise less readily. The biggest risk with solid materials is when they occur in sheet or foam form. In each case oxygen access is greater and when heat is applied they become hot much more rapidly than large masses of material.

Evaluating fire hazard is a highly complex process and this is reflected in practice by the numerous BS tests for the different aspects of fire. The chief hazards can be summarised:

- Heat itself causes burns.
- Fire may endanger the structure.
- Many materials generate toxic fumes when they are heated (even some which do not readily flame or spread fire). This is a common cause of fatalities in house fires in which fumes may affect the occupants – especially while asleep at night – well before there is exposure to heat.
- Materials often generate smoke, which makes breathing difficult and tends to cause panic and disorientation as people try to escape.

Building and Fire Regulations are chiefly concerned with **escape**. The welfare of the structure itself is of secondary importance, though insurance companies will be anxious to minimise **building damage** in addition and the **fire services** will also be concerned about provisions in the building for tackling fires effectively without **undue risk** to personnel.

Illustrations of some current British Standards are given in Fig. 1.12.

Fig. 1.12 *British Standards relating to fire resistance.*

Surface spread of flame
[BS 476: Part 7]

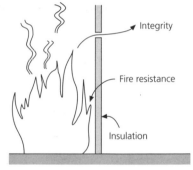

Fire resistance, integrity and insulation
(door/partition)
[BS 476: Parts 20–24]

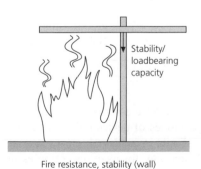

Fire resistance, stability (wall)
[BS 476: Parts 20–24]

External fire exposure test
[BS 476: Part 3]

Durability

> *A material may be said to be durable in any one situation if it fulfils all its performance requirements, either for the planned lifetime of the building, or for a shorter defined period where this is acceptable.*

Service lives for components of less than the building life may be appropriate where replacement is straightforward, where there is a severe cost penalty of longer life or where replacement arising from changing user requirements may be desirable.

It is often very difficult to predict the durability of individual components. Also, since failures in a very small proportion of the items in use may be unacceptable, the only safe course of action may be to 'over-design' them so that materials in the worst likely situation should be satisfactory. In consequence, many buildings last much longer than their design lifetime. Table 1.2 summarises some of the main modes of deterioration of major materials groups.

Table 1.2 *Modes of deterioration of the major materials groups*

Type	Form of deterioration	Recommendation
Metals	Surface deterioration by electrolytic corrosion in damp conditions	Use resistant metal such as stainless steel or apply coating to exclude moisture
Bricks, stone, concrete (porous)	Moisture penetration followed by frost or chemical action. Deterioration at or below surface.	Use non-permeable forms; weatherings to shed water; impermeable coatings to prevent water penetration
Timber and timber products	Fungal attack (damp conditions). Insect attack (normal conditions)	Exclude water; apply preservatives
Polymeric materials	Ultraviolet degradation	Protect from sun – e.g. carbon black pigment, or use indoors/underground. Use stabilisers or surface coatings

It will be noted that the majority of modes are associated with the effects of water. A further important point is that if these modes of deterioration are avoided, as should be the case with good practice, long life should be achieved; most materials do not 'age' except by these specific mechanisms. There are, for example, samples of timber, concrete and metals over 1000 years old and in good condition.

Safety in use and health

Public awareness of health and safety issues is increasing rapidly and the construction industry has come under scrutiny both from the point of view of safety of the construction team and the general public during the construction process and because the finished building might present a hazard to the occupants. A detailed examination of all the aspects of health and safety relating to building materials is beyond the scope of this book, though Table 1.3 indicates some of the possible safety hazards posed by materials, together with recommendations for addressing them.

Table 1.3 *Common construction industry materials: risks*

Substance	Situation	Risk	Remedy
Lead	Formerly in paints, pipes, solders	Health risk if ingested, especially to children	Remove existing lead pipework. Specify lead-free paints and solders
Radon gas	Diffuses from some granite rocks through ground floor	Radioactive gas	Isolate ground beneath buildings
Formaldehyde gas	Present in some foams (e.g. cavity fills) and wood products	Can cause nausea	Adequate ventilation after installation
Chorofluoro-carbons (CFCs)	Used in refrigerants, air conditioning systems, propellants for aerosols and foaming agents for some plastics	Depletes ozone layer, hence contributes to global warming and increases UV radiation at earth's surface	Check specifications for products which might contain CFCs. Less harmful substitutes now available
Wood preservatives	During storage, transportation and at or soon after application	Risk to human, animal or plant life	Use only in safe situations under strict control. Consider other ways of controlling risk
Asbestos	Not currently used but may be present as insulation or other forms especially in older buildings	Risk of lung cancer	Arrange for safe removal/disposal if found
Polyvinyl chloride (PVC)	Widely used for waste systems, windows, miscellaneous applications. Plasticised version in fabrics	Chief risk is in manufacture, where choride pollution in effluent is a risk. Small risk of toxic gas in fire	Stringent precautions during manufacture. Normal fire precautions
Solvent-based paints	Factory and site	Risk to operator and of general atmospheric pollution	Use water-based paints where possible
Bitumens	Site application	Vapours of hot applied material may be hazardous	Take precautions during use

Environmental issues

There is currently relatively little legislation concerning environmental issues and therefore a strong tendency for these matters to be overshadowed by economic or other criteria when selecting materials for building.

Environmental matters are beginning to feature much more strongly in the assessment of building materials and as this trend is likely to continue to gather momentum into the new millennium, the following key aspects of environmental performance will be considered:

- Embodied energy
- Recycling potential
- Environmental management

In practice other factors such as potential damage to the landscape caused by the extraction and disposal processes should also be considered.

Embodied energy

Energy should be regarded as a precious and finite resource and although it is relatively cheap at the present time there is an economic advantage of use of materials which take less energy to convert them from raw materials into the final, in-situ product. In seeking new materials, emphasis should be given to those materials which consume less energy in their manufacturing process, and such materials should become increasingly competitive when, as is very likely, energy costs rise. The process can be quite complex since costs must be assessed at **each stage** as shown below.

Table 1.4 shows qualitatively the energy inputs associated with major groups of building materials. Such a table can only be approximate and the transport cost in particular will depend upon the geographic proximity of the various stages from extraction to point of use. It does, however, permit general conclusions to be reached. It should also be appreciated that the energy used to produce materials for construction is in general considerably smaller than that used to operate the building over its lifetime. The biggest component of energy input for all applications – industrial, public and private – is usually heating and attention needs to be given to how the need for such inputs can be reduced. This is one illustration of the need to consider the lifetime scenario relating to any one material, rather than simply the issues concerned with its production.

Recycling potential

This may conveniently be described as a hierarchy of Rs, as shown in Table 1.5. Only when none of these is feasible should **disposal** be considered. Materials/situations in which disposal is the only practical option should be regarded as failures in an environmental sense.

Recycling may have various goals:

- Reduction of raw materials consumption
- Reduction of fuel consumption in manufacture
- Reduction in waste generation

In any one context the aims of recycling should be established and the feasibility of recycling examined bearing in mind the following:

Table 1.4 *Major groups of building materials: embodied energies*

Material	Extraction	Manufacture	Transportation	Cost in use	Comments
Timber	Low: felling and transport	Very low: machining/ seasoning	Low: timber yard to site	Generally low: requires occasional protection	Timber overall scores highly
Concrete	Low: aggregate extraction	Medium: requires 10–20% of cement – a high-energy material	Fairly high due to high density	Low: provided material is durable	Overall moderate energy cost
Clay bricks	Low	High: bricks must be heated to 1100 °C	Fairly high due to high density	Very low: provided material is durable	Overall high energy cost
Plastics	Medium: obtained from oil	High: complex manufacturing process	Low: low-density material	Medium: usually limited life	High energy cost, offset by advantages for common applications
Steel	High: ore must be obtained and transported	Very high: steel is molten during manufacture	High: high-density material	Low: provided rust protection applied	A high-energy material, with significant structural advantage
Aluminium	High: ore must be obtained and transported	Very high: high electricity cost in process used	Medium	Very low: usually protected by oxide coating	A high-energy material: advantages are durability and lightness

- Energy costs involved with recycling comparative to energy savings in 'primary' manufacture.
- Economic cost of recycling, including recovery of material for the process.
- Possible environmental implications of the recycling process itself, including possible toxic emissions/residues.

Environmental management

There are calls for sustainable construction, which is a subset of a sustainable society. Such a society should be able to continue to operate without compromising prospects

Table 1.5 *Recycling options*

Option	Comments
Reduce	Problems can be avoided at source if wastage in production is reduced and materials have inherently low maintenance characteristics
Reuse	This implies minimum recovery/reprocessing cost – for example, roofing tiles and brick bedded in a lime mortar
Recycle	Some manufacturing is involved – for example, lead, steel, glass. Solid wood could be reprocessed into chipboard
Recover energy	This may still be worthwhile if combustible materials such as timber, plastics and rubber are incinerated and the heat recovered, say, for district heating schemes
Disposal	Should be the last resort

for operation of future societies. In environmental terms, this means avoiding the following:

- Depletion of finite resources.
- Adversely affecting the environment by pollution or waste.
- Adversely affecting the environment by energy emission.
- Adversely affecting the environment in a broader sense, such as upsetting the eco-balance of wildlife.

At the present time these criteria are rarely fulfilled and there is an urgent need to address issues, especially where it is believed that measurable adverse changes are already taking place. In addition to the BS ISO 9000 Quality Management Systems described above, many organisations are now operating Environmental Management Systems in which environmental implications of all operations are systematically considered and controlled. The ISO version of this is BS EN ISO 14001 and the corresponding European Standard is the 'Ecomanagement and Auditing Scheme'. Such schemes may be driven by local authority initiative or, in the case of for example, materials suppliers, by public perception and demand.

SELF-ASSESSMENT QUESTIONS

1. Distinguish between the functions of British Standards and Agrement certificates.

2. Describe the use of the CE mark in relation to construction products.

3. Explain the difference between the terms 'quality control' and 'quality management systems'.

4. Explain the difference between the terms 'strength' and 'stiffness'. Give examples of materials of

(a) high strength and high stiffness;
(b) low strength and high stiffness;
(c) high strength and low stiffness;
(d) low strength and low stiffness.

5. Explain the difficulty of measuring the strength of a material
   (a) in compression;
   (b) in tension.

6. A material is to be used to clad the entrance hall to a large public building. Indicate three aspects of its fire performance which should be considered.

7. Compare the embodied energies of
   (a) timber;
   (b) concrete;
   (c) steel.

8. Describe the options available other than disposal at the demolition stage of a building.

STANDARDS

BS 476: Part 3: 1975: *External fire exposure test*.
   Part 4: 1970 (1984): *Non-combustibility test for materials*.
   Part 7: 1987: *Method of classification of the surface spread of flame for products*.
   Part 11, 1982 (1988) *Method for assessing the heat emission from building materials*
   Parts 20–22: *Determination of fire resistance*
   Part 31: *Methods for measuring smoke penetration through doorsets and shutter assemblies*
   Part 31.1: 1983: *Method of measurement under ambient temperature conditions*
BS ISO 9000: *Quality management and quality assurance standards*
BS EN ISO 14001: *Environmental management systems. Specification with guidance for use*

2 Structure and performance of materials

Chapter summary

- The atom.
- Electron shells.
- Atomic bonding.
- Bond strengths.
- More recent materials.

2.1 Introduction

The materials available for use in construction and the ways in which they interact with each other are very much dependent upon their chemical make-up, which depends in turn upon the characteristics of the atoms forming them, so a brief introduction to this topic will be given.

2.2 The atom

The atom is the smallest particle into which materials can be divided by ordinary (chemical) means. Since subatomic reactions are rare in everyday events, the balance of distribution of atoms which exists in the earth's crust at the moment will not change on any measurable timescale so that relative abundances of the various atom types (or elements) are, for all practical purposes, fixed.

All atoms comprise two 'sections', namely a **nucleus** and one or more **electrons** (the electron shell). Each of these sections has important implications for the availability and interaction of materials with each other.

Nucleus

This may be regarded as the 'massive' part of the atom and comprises protons and neutrons. These particles are extremely small, protons and neutrons each having a mass of about 2×10^{-27} kg. The proton also has a positive charge (about 10^{-19} of a coulomb, which is the charge flowing when a current of one amp flows for one second). The

Table 2.1 *Subatomic particles: relative masses and charges*		
Particle	*Relative mass*	*Relative charge*
Proton	1	+1
Neutron	1	0
Electron	0.0005	−1

Fig. 2.1 *Representation of electron orbits in relation to nucleus.*

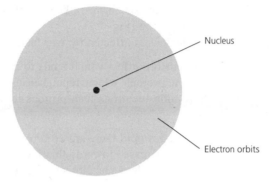

Fig. 2.2 *The alpha-particle comprising two protons (positive charge) and two neutrons (no charge).*

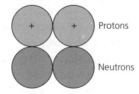

charges and masses of these particles are compared with those of the electron (to be discussed later) in Table 2.1.

While it would be wrong to regard nuclear particles as completely motionless, they nevertheless occupy an extremely small space compared with that occupied by the atom as a whole, so most of the atom could be regarded as empty space (Fig. 2.1).

The protons and neutrons are held together by very short range forces which disappear at separations greater than about 10^{-15} m. The forces are peculiar to nuclear particles – they are neither electrostatic nor gravitational in origin. The stability of nuclei stems from the fact that, when protons and neutrons come together, there is a small release of mass – the basic principle of nuclear energy. A particularly stable arrangement is a unit comprising two protons and two neutrons – called the **alpha-particle** (Fig. 2.2). (When atoms disintegrate they often do so by emitting an alpha-particle.)

The number of protons in a nucleus determines the element type since the number of protons gives each atom a unique 'fingerprint' and has important implications for every

Fig. 2.3 *Hydrogen atom: one electron is in orbit around one proton; the atom is mainly empty space.*

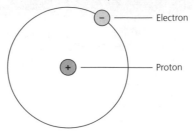

aspect of behaviour of materials formed. The simplest of all materials is hydrogen, which contains just one proton (Fig. 2.3).

The stability of a nucleus is affected by two factors:

- A sort of 'surface tension' effect which tends to make larger atoms more stable.
- A repulsion between protons which repel because they are like charges. Larger atoms tend to have more neutrons than protons in order to help minimise proton–proton repulsion.

The first effect dominates until there are 26 protons in the atom, and nuclei become more stable as their size increases towards this level. For this reason element no. 26, iron, is the most stable of all elements. Not surprisingly, iron is abundant in the earth's crust (although it is not the commonest element).

At larger nuclear sizes, the greater number of protons repel each other and to counter this effect larger nuclei tend to have greater numbers of neutrons than protons. Lead, with 82 protons, has 124 neutrons and is the largest stable element. Elements of larger nuclear size than lead are described as **radioactive** since they disintegrate (usually slowly) to form lighter elements. As indicated above, this process often involves the emission of an alpha-particle as well as short wave radiation similar to X rays, which are a health hazard. The practical range of elements that can be used in construction is therefore 82 and within this range some elements are much more abundant than others. The abundance of elements is a major factor in determining the cost of construction materials. Figure 2.4 indicates the nine most common elements. Oxygen is the most common, and most of the solid materials in the earth's crust are chemically combined with it. Silica or silicon oxide (SiO_2), the main constituent of sand, is the most common compound. Calcium, which follows iron in terms of abundance, is a very important element for construction. It plays a key role in the action of Portland and high-alumina cements, hydrated lime and gypsum plaster. The elements shown comprise 99 per cent of the total. They are followed (in order) by:

- Titanium
- Phosphorus
- Manganese
- Nickel
- Sulphur
- Cobalt
- Chromium

Fig. 2.4 *Relative abundances of nine most common elements in the earth's crust.*

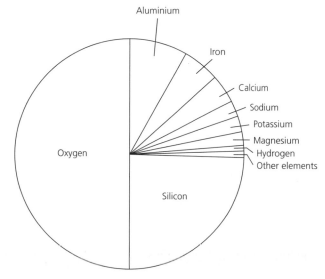

These make up most of the remaining 1 per cent, but there are more than 60 others that occur in very small amounts.

It will be noticed that carbon, the basis of all living organisms as well as being the backbone of most plastics, is relatively rare, largely because there is very little carbon below ground. The reserves of coal, oil, carbonate rocks and other materials are derived from earlier life. There are, however, substantial quantities in the atmosphere, which includes approximately 0.03 per cent carbon dioxide by volume. This sustains all vegetable life and therefore, indirectly, all animal and human life.

In order to redress the balance of elements in the earth's crust it might be considered desirable to change or 'engineer' elements around, converting less useful ones to more useful ones. While it is possible to change one element into another by nuclear physics, the energies involved are very high – the same energy that is involved in nuclear power except in the reverse direction. This could never be carried out on a commercial scale and for construction purposes we must make the best use of elements in their existing proportions.

It should be emphasised that while the nucleus determines the character of the atom, and therefore its chemical behaviour, it is not directly involved in the chemical changes that materials undergo during their production, service life and eventual decay. All these changes are linked to behaviour of the outer electron shells.

Electron shells

The description of the electron is given in Table 2.1. Note that it is very light relative to nuclear particles and that it has a negative charge which causes it to be attracted to the nucleus. The electrons orbit around the nucleus rather like a satellite orbits the earth, although the analogy is not perfect because the nucleus must be regarded as much more

Fig. 2.5 *Representation of calcium – it has the following shell occupation: 2, 8, 8, 2.*

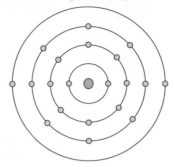

densely packed than the earth, the atom as a whole being mainly empty space. It is also not always correct to think of the electron as a particle; it is sometimes more convenient to regard it as a wave form.

An important requirement for a stable atom is that the electron charge must balance the nuclear charge, so there will be equal numbers of electrons and protons in a stable (uncharged) atom.

A further important requirement is that when atoms have several electrons in orbit, the orbits must, in some way differ. If they were identical there would be a risk of electrons 'colliding' with each other. This requirement is fulfilled by assigning energies to orbits, the lowest energy orbit normally being the one that fills first. This leads to the concept of electron shells, each shell containing a permitted number of orbits. The orbits have been derived by studies of quantum mechanics but it will be sufficient here to simply list the orbit patterns available according to atomic size. The orbits follow a simple mathematical sequence:

Shell 1	2			
Shell 2	2	6		
Shell 3	2	6	10	
Shell 4	2	6	10	14
etc.				

It should be noted that within each group no two orbits are identical. The energies of the orbits increase both downwards and to the right. Since the 10 and 14 electron sections of each group are actually of higher energy than the 2 and 6 electron sections of the following shell, we find that the outermost shell of any atom always contains between zero and eight electrons. For example, calcium (Ca) has the electron arrangement shown in Fig. 2.5. This principle gives rise to the Periodic Table of elements (simplified) shown in Fig. 2.6.

Much as the nucleus has a 'binding energy' which represents the energy required to break it up in some way, so the electron shells have similar energies – the energy needed to extract an electron from the atom. This is a maximum when each shell contains eight electrons since the addition of another electron would require the next shell to start filling. (The exception to this is the first shell containing hydrogen (H) and helium (He) since the innermost shell can only accommodate two electrons.)

Fig. 2.6 *Simplified Periodic Table.*

Shell capacities are:
Shell 1: 2 electrons
Shell 2: 2 + 6 electrons
Shell 3: 2 + 6 + 10 electrons
Shell 4: 2 + 6 + 10 + 14 electrons, etc.

Groups

Shell number	1	2		3	4	5	6	7	8
1	1 H								2 He
2	3 Li	4 Be		5 B	6 C	7 N	8 O	9 F	10 Ne
3	11 Na	12 Mg		13 Al	14 Si	15 P	16 S	17 Cl	18 A
4	19 K	20 Ca	21–30[a]	31 Ga	32 Ge	33 As	34 Se	35 Br	36 Kr
5	37 Rb	38 Sr	39–48[b]	49 In	50 Sn	51 Sb	52 Te	53 I	54 Xe
6	55 Cs	56 Ba	57–80[c]	81 Tl	82 Pb	83 Bi	84 Po	85 At	86 Rn

[a] Elements 21 to 30: these are the 10 transition elements; comprising scandium, titanium, vanadium, chromium, manganese, iron, cobalt, nickel, copper, zinc. Their properties are rather similar because they correspond to electrons filling orbits in shell 3, shell 4 at this stage having two electrons in it, which largely dictate their character.

[b] Elements 39 to 48: these represent the 10 electron orbits in shell 4 filling in a similar way to the 10 elements as in note a, though the elements are less abundant.

[c] Elements 57 to 80: these represent a further 14 electrons filling orbits in shell 4 plus 10 electrons filling shell 5.

Elements to the left of the bold line form metals when pure

Shells containing eight electrons (plus, of course, helium) give rise to **inert gases** – that is, elements of maximum chemical stability. They have the advantage that they are reluctant to become involved in any chemical reaction, though they have the related problem that, because of their stability, they do not readily bond either with themselves or with any other material. Their uses are therefore limited and specialised. They are:

Atomic number	Element
2	Helium (He)
10	Neon (Ne)
18	Argon (Ar)
36	Krypton (Kr)
54	Xenon (Xe)
86	Radon (Rn)

The corollary to this is that all other atoms 'seek' a full electron shell and this is usually achieved by the process of atomic (chemical) bonding.

A further important point concerning the periodic table is that larger elements tend to be more dense on account of their highly positively charged nuclei which attracts electrons to them. It is no coincidence that lead, element number 82 and the largest stable atom, has a relatively high density.

2.3 Atomic bonding

The concept of atomic bonding is crucial to construction materials because without it there would be no solid or liquid state; the universe would consist of an enormous gaseous soup! The rigidity of steel, the abrasion resistance of granite, the tensile strength of carbon fibres, the temperature resistance of tungsten and the elasticity of rubber are all directly attributable to their bond structures. A basic familiarity of bond types will aid understanding of what can be expected of the various materials groups and how they can work together to give functional and durable buildings. When atoms are linked by a bond the combination is known as a 'molecule'. Molecules may comprise as little as two or many thousands of atoms. The latter are much more likely to produce a solid, rigid material.

- From the atom's point of view, as explained above, bonding brings about a reduction in the energy of the material. From the user's point of view bonding brings about increasing rigidity.
- From the atom's point of view 'ideal' bonding results in each atom effectively having a full (or empty) electron shell. From the user's point of view such bonds will be the strongest, though in order to produce strong solids, such bonding should be extended throughout the sample of material and this is often very difficult to achieve.

The most important bond types are discussed below.

Primary Bonds

In primary bonding the atoms largely fulfil the empty/full shell requirement. They comprise ionic, covalent and metallic bonding. All these bonds involve outer electrons only. There is no bonding potential in inner shells since electron arrangements in these are stable.

Ionic bond

This occurs when, for example, an atom containing a single outer electron encounters another atom with a single electron missing in its outer shell. The electron in the first atom transfers to orbit the nucleus of the second atom. Hence, they have empty and complete shells, respectively, but the previously uncharged atoms will now be charged and, therefore, attract one another, coming together until this attraction is balanced by the repulsion of their respective electron clouds. These charged atoms are known as 'ions'. If further atoms of each type are available, the process will continue and a crystal will form. Crystalline materials, rather than small combinations of atoms, offer much better prospects for use as construction materials because, as indicated above, the resulting material is more likely to exist as a solid. Common salt (NaCl) is an example (Fig. 2.7). Sodium has a single outer electron and chlorine has seven outer electrons. Similar reactions will take place between elements with an excess or shortage of two electrons; for example, calcium and oxygen (CaO). By combination in appropriate proportions molecules such as, for example, calcium chloride ($CaCl_2$) can be formed.

Fig. 2.7 *Ionic bonding between sodium and chlorine forming common salt.*

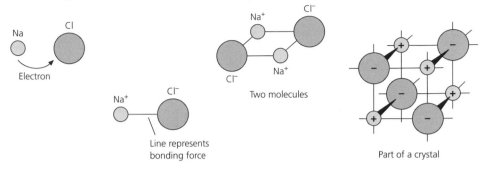

The ability of ionic materials to form crystals depends upon the relative sizes of the ions; where they differ greatly, packing problems will prevent crystals from forming. The presence of ions in water may interfere with the force between the ions of crystals so that many ionic compounds 'dissolve' in water, producing a dispersion of ions – for example, common salt. Water in smaller quantities may actually assist in crystal formation since the water molecules become involved in the crystal structure – for example, set gypsum plaster. Cements also set and harden by the slow dissolution and crystallisation of calcium silicates in the presence of water.

Conversely, some salts in brickwork, for example, crystallise if damp, taking water into their bond structure. This may lead to an expansion and cause extensive damage to porous building materials.

Ionic materials are characterised by the presence of small electrostatic charges and these may assist in the formation of secondary bonds as explained below.

Covalent bond

This type of bond can be formed between atoms which do not have suitable arrangements of electrons for ionic bonds to be formed – in other words, similar deficiencies or excesses of the bonding atoms. If atoms come together so that one or more of their outer electrons orbit both nuclei, each atom has a stable octet for a part of the time. In methane, for example, four hydrogen atoms, each having one electron in its shell, approach a carbon atom having four electrons in its outer shell and each hydrogen electron encompasses both its own nucleus and the carbon nucleus. In return, the four electrons from the carbon atom each encompass a hydrogen atom. The arrangement is shown schematically in Fig. 2.8. The carbon–hydrogen bond is, in fact, very versatile, forming the basis of many organic materials. Some important examples are shown in Fig. 2.9. Note that it is possible for two or three electrons to orbit both atoms, as in ethylene (ethene, C_2H_4) and acetylene (ethyne, C_2H_2), forming double and triple bonds, respectively. In both natural and synthetic organic materials, the carbon may form long-chain molecules containing many thousands of atoms. The covalent bond may give rise to fibrous materials. In contrast, many materials comprise very small molecules bonded in this way – for example, oxygen (O_2), nitrogen (N_2) and carbon dioxide (CO_2).

Fig. 2.8 *Sharing of carbon and hydrogen electrons to produce methane.*

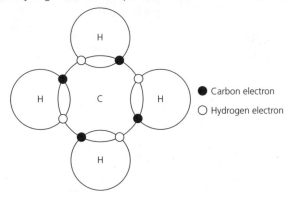

● Carbon electron
○ Hydrogen electron

Fig. 2.9 *Various ways of representing the structure of some organic compounds (i.e. formed from carbon). Note that, in each compound, hydrogen atoms attain two electrons and carbon atoms attain eight electrons.*

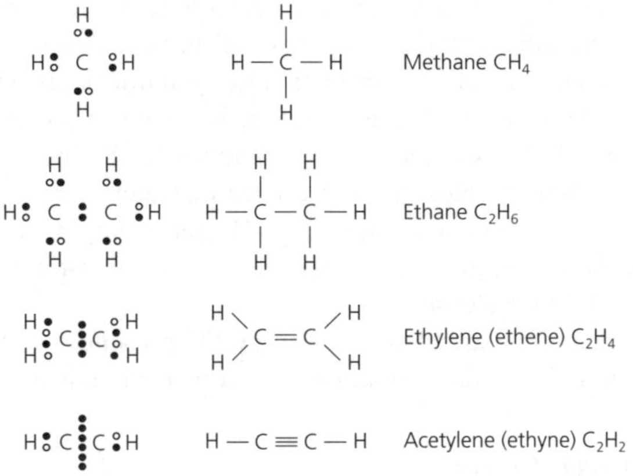

Methane CH_4

Ethane C_2H_6

Ethylene (ethene) C_2H_4

Acetylene (ethyne) C_2H_2

Another common feature of the covalent bond is that, unlike the pure ionic bond, bond directions are well defined, because orbits of bonding electrons must fit in with those of non-bonding electrons. This can lead to great rigidity of which, perhaps, the most notable case is diamond.

Covalent bonds are also less likely to be affected by water since they are less dependent upon the formation of ions by atoms.

It would be misleading to suggest that all bonds are either ionic or covalent – many bonds are a combination of the two. For example, anhydrous calcium sulphate (the basis of gypsum plaster) consists of a sulphate ion (SO_4^{--}) and a calcium ion (Ca^{++}). The sulphur (six outer electrons) shares two of its electrons with each of three oxygen atoms (each with six outer electrons), the two extra electrons which are necessary being borrowed ionically from the calcium which forms the Ca^{++} ion, as in Fig. 2.10. Note that the sharing between the sulphur and the oxygen is in one direction only since, with the two ionically borrowed electrons, the sulphur has eight electrons and has no need to borrow from the oxygen atoms. Each oxygen is, on the other hand, two electrons short and must, therefore, share with the sulphur.

Fig. 2.10 *Ionic/covalent bonding producing calcium sulphate.*

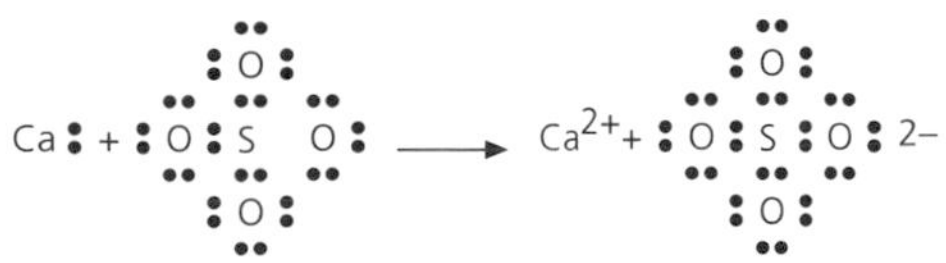

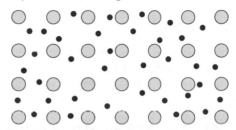

Unstable   Stable

Fig. 2.11 *Electrons in a 'cloud' around positive ions, forming the metallic bond.*

Metallic bond

When atoms of a single type and containing a few outer electrons approach one another, the electrons' orbits may change to a nondescript nature – they move around the nuclei rather like a gas and cause an overall attraction between the positively charged atoms and the electrons themselves (Fig. 2.11). Such bonding is clearly non-directional and the atoms therefore pack in tight patterns, resulting in crystals. An electric potential will cause a drift of the electron gas through the metal, constituting an electric current, hence, metals conduct electricity. The thermal conductivity of metals is also high. The electrons are mobile and readily carry thermal energy from one point to another by collision.

The great majority of elements form metals when they are present in their pure state; all those to the left of the bold lines of Fig. 2.6 being metallic when pure. This is not so much an indication of the stability of metals as a result of the fact that the electron configurations of these elements in the pure state do not readily permit formation of ionic and covalent bonds. Indeed, if oxygen is present, many of these combine quickly with it in ordinary atmospheres, producing more stable covalent or ionic bonds. The occurrence of most of these elements as oxides or other compounds is an indication of their instability in metallic form. Tin (Sn), element number 50, can in fact change to a covalent form simply by cooling, hence, it reverts slowly to a grey powder below 18 °C. There is probably some tendency to form covalent bonds in most metals. According to the extent of this, three different metallic crystal types may be formed. The influence is noticeable in steel in which covalent tendencies, at any temperature below red heat, lead to a less 'metallic' crystal structure. In all cases, however, the tight packing of atoms results in materials which are of much higher density in metallic form than their non-metallic counterparts. Although the metallic bond is not, in general, as strong as ionic and covalent bonds, the high density of packing of ions in metal crystals results in metals being a most important bulk structural material.

Fig. 2.12 *Bonding of polyethylene (polythene) chains by van der Waals forces.*

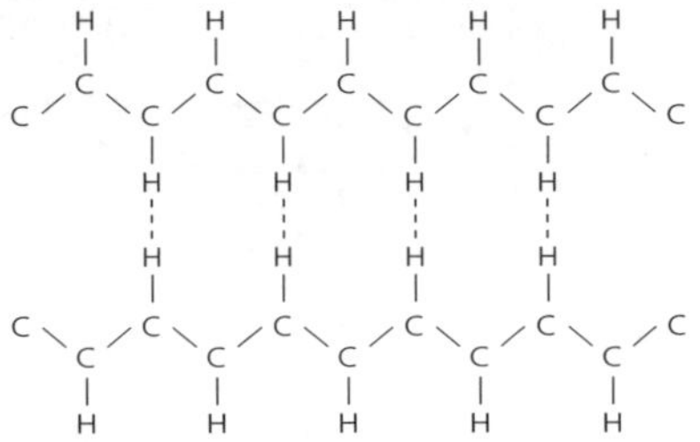

Secondary bonds

In the bonds considered so far, the atoms have combined to produce a more stable arrangement by interaction of their electrons. In some cases, the materials produced are crystalline and, therefore, solid. In others, a stable arrangement is produced by formation of small groups of atoms – for example, oxygen (O_2) or ethylene (C_2H_4). It is commonly known, however, that the latter materials may be liquefied or even solidified if the temperature is lowered sufficiently. This can be explained by 'van der Waals' bonds. In any one molecule, the electrons in orbit around it produce small instantaneous eccentricities of charge which induce similar charges in neighbouring molecules causing them to be attracted to it. Hence, this type of bond is operative between any adjacent molecules and, although it would, on average, have a strength of only about 1 per cent of ionic/covalent bonds and is easily overcome by thermal energy, it may produce a solid with a certain amount of strength, especially in the more 'polar' molecules. The softer plastics are examples of molecular chains made solid in this way (Fig. 2.12). Van der Waals bonds also are important in concrete and may be used to explain the phenomenon of creep. Colloidal properties of materials such as clay are similarly explained. Van der Waals bonds are responsible for the solid nature of graphite – they cause the attraction between covalent bonded carbon sheets. If materials such as timber, bitumen, paints and adhesives are added to the list, the importance of the bond will be appreciated.

It may be considered that the bonds, which will solidify almost all non-crystalline, micro-crystalline or fibrous materials, manifest themselves in surprisingly different ways in the materials listed. The differences are, however, not as great as may at first be supposed. Essentially, there are two groups: those which absorb water (e.g. clay, concrete and, perhaps, timber) and the remainder, which are impervious to water. When water is absorbed into porous materials, it tends to cause swelling by being attracted to solid surfaces, thus reducing the effectiveness of the van der Waals bonds. For this reason, there may be an initial increase in the strength of clay, concrete and timber if water is removed by heating. This is, however, followed by a decrease of strength at higher temperatures. Materials in the remaining group all lose strength immediately on heating.

2.4 Bond strengths and strengths in practice

It should be appreciated that the strength of bulk materials will be determined by the weakest bonds in a particular sample of a material – the 'weak link in the chain' principle applies. Bond strengths depend on a number of factors, but orders of magnitude of tensile strengths corresponding to materials fully and perfectly bonded by the various bond types is:

Ionic/covalent/metallic $10,000–40,000 \text{ N/mm}^2$
Secondary $5–20 \text{ N/mm}^2$

The first figure given may seem extremely high when compared with bulk strengths of even high-quality material such as high-tensile steel (tensile strength around 2000 N/mm^2) but this is because bond strengths are rarely translatable into practical working strengths.

It should be remembered that if strengths such as those above are to be obtained in practice, the bonding must be extended on a very large scale, ideally throughout the piece of material to be used. Although this is almost always impossible, the following are ways in which extended bonding *might* take place. In each case the limitations to size are considered.

Crystals

Crystals are regular arrays of atoms or molecules. Theoretically any small group of atoms can form crystals. An illustration of ionic crystals was shown in Fig. 2.7. It is found, however, that when bonds form on manufacture of a material, crystals grow from many different nuclei in the material. Where they meet there will be regions of disorder and poor bonding which severely limit the performance of the resulting solid. In addition, crystals may contain defects, some of which are discussed in Chapter 7. Diamonds are examples of a perfect crystal and illustrate the great strength that can be obtained, but it is well known that they are very difficult to grow to commercial sizes and offer no prospects for use as construction materials.

Crystals do not have to be three dimensional; they can form in two dimensions (sheets) or one dimension only (fibres). In each case there will be strong bonding only within the crystal, therefore sheets and fibres do not necessarily result in very strong materials.

Examples of sheet-like crystals are graphite and many minerals such as mica. Some clays contain minerals in sheet-like form. Clay when damp has a slippery feel characteristic of sheet-type particles, while graphite is used as a solid lubricant for the same reason.

Examples of one-dimensional bonding are found in fibrous materials such as timber, which contains partially crystallised polymer chains comprising carbon, hydrogen and oxygen.

Further examples of fibrous materials based upon molecular chains are asbestos (comprising silicate chains) and organic fibres (comprising carbon chains).

Amorphous materials

These are materials which do not crystallise. They are sometimes referred to as 'glasses' because, if fully amorphous, light may travel through the solid formed (in the absence or crystal boundaries which tend to scatter light). Amorphous products result when a wide range of materials is heated until fusion (melting) takes place. Glasses are particularly likely to form when the material has a complex or mixed composition, since crystal growth requires ordered molecular arrangements. Examples of glasses are:

- Silica-based glasses used for glazing.
- Many thermoplastics materials, such as PVC.
- The fabric of clay bricks (though they do not appear glassy because the material is only partially fused).
- Some stones, such as flint.

Performance of real materials

There are no common materials in which primary bonding extends throughout the material. Even if they did exist it is doubtful whether full use of structural potential could be made since very strong materials do not usually have greatly increased E values; they therefore tend to operate at high strain levels which may not be satisfactory for construction materials. In most materials applications in construction, there is a need for rigidity.

Table 2.2 shows examples of the main construction materials, their main bond types and the factors limiting their strength.

Table 2.2 *Common construction materials: factors limiting strength*

Material	Bond type	Factor limiting the strength
Metals	Metallic crystals	Dislocations; grain boundaries (at higher temperatures, see Chapter 7)
Natural stones	Crystalline (granite)	Grain boundaries; cracks
Natural stones/glass	Amorphous (e.g. flint)	Inclusions (impurities); cracks
Cements	Microscopic ionic/covalent interlocking fine crystals	Non crystalline areas; cracks; pores (voids)
Timber	Fibrous crystalline composite with fibre interlock/ secondary bonding	Buckling of fibres; slip between fibres; secondary bonding forces
Thermosets	3D amorphous ionic/ covalent molecules	Limited numbers of bonds
Thermoplastics	Polymer chains with interlocking/van der Waals forces	Van der Waals forces

2.5 More recent materials

A consideration of how materials could be improved follows naturally from Table 2.2. Within the metals group, it would be fair comment that high-performance metals already exist – for example, steel wire is available with strengths approaching 2000 N/mm^2. The main problem with metals is:

- High manufacturing energy requirement
- High density
- High cost

Metals undoubtedly fulfil a vital function in construction, but in view of the above, use should be restricted to situations where requirements cannot be met by other materials.

In terms of other materials groups, weaknesses could be broadly classified as:

- Poor tensile strength and impact properties of materials such as concrete.
- Low stiffness of plastics.

Progress has been made in addressing these deficiencies. Two techniques will be briefly described.

High-performance concretes

Attempts to improve the tensile strength of concretes must focus upon the main cause of weakness. Cement hydrate contains pores and most concretes also contain pores resulting from the evaporation of excess water after hardening.

Efforts have been made to control both the total porosity of the material and the size of pores.

Control of porosity

It is well known that reducing the water/cement ratio of concrete (see Chapter 3) increases strength since

1. Unhydrated cement is stronger than hydrated material.
2. Less water in the mix results in fewer pores in the long term.

Very high strengths (compared with typical design strengths) can be obtained by use of very low water/cement ratios. These concretes are, however, very difficult to handle in the fresh state. Additives such as super plasticisers and microsilica can permit concrete compaction to be achieved, albeit with difficulty. Concretes with compressive strengths of over 100 N/mm^2 and tensile strengths of 15 N/mm^2 can be obtained in this way, but such products are best suited to factory production. They would permit a reduction in section sizes but would still need to be reinforced in situations of high-tensile stress or impact.

Reduction of size of pores: MDF cement

Experiments have been carried out to find ways of reducing the size of pores since it is known that larger pores are mainly responsible for strength reduction.

The origin of the larger pores or defects appears to be packing problems in cement particles caused by friction between them. It is now possible to eliminate these in certain special Portland and high-alumina cements by the addition of water-soluble

Fig. 2.13 *Some simple artifacts made from MDF cement. (Reproduced courtesy of University of Oxford, Department of Metallurgy and Science of Materials)*

polymers, such as polyvinyl alcohol/acetate, which reduce particle friction, allowing defects to be removed by rolling low water/cement ratio mixes into a dough-like material until macro-defects are no larger than 0.1 mm. Ordinary plasticisers and super-plasticisers cannot be used for this process because they do not appear to reduce particle friction in such mixes.

Residual porosity can be removed by heating the moulded material to 80 °C for a short time. This causes a shrinkage of almost 10 per cent in the material as it hydrates, corresponding to a large reduction in porosity. The final product has a flexural strength of around 150 N/mm^2 with an elastic modulus of about 50 kN/mm^2, indicating a strain capability of around 3000×10^{-6}; rather larger than the yield strain of mild steel.

Figure 2.13 shows examples of small artefacts produced from MDF. Widespread use of such products has yet to occur, partly because properties deteriorate on wetting.

Fibrous composites

Fibres may address the shortcomings in cement-based materials and in plastics as they have the distinct structural advantage of being of small diameter. It has been explained above that practical strengths are often limited by defects in materials. It is found that fibres have, in general, many fewer defects than bulk equivalents, due at least in part to the fact that stresses in production (for example, thermal stresses) are greatly reduced and also that there is less scope for large defects in samples of small size. The tensile strength of glass fibre is increased, for example, from about 100 N/mm^2 to around 2000 N/mm^2 by reducing the size of individual filaments to about 10 μm. (It should be noted that such glass fibres are much smaller than those designed for thermal insulation.)

Fig. 2.14 *Use of fibres to bond cracks together.*

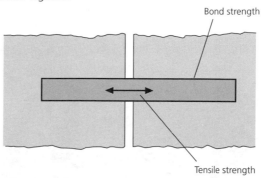

Bond strength

Tensile strength

Table 2.3 *Cementitious materials: using fibres to improve properties*

Fibre volumes used are quite small

Material	Fibre Type	Proportion by volume	E value (kN/mm²)	Strength (N/mm²)	Application
Cement	Glass (special alkali-resistant type)	5%	70	2000	Unreinforced cladding panels
Concrete	Steel	10%	210	1000	Stress transfer between precast floor units
Concrete	Polypropylene	0.2%	8	400	Crack control in floors; impact resistance in precast units
Gypsum	Glass	4%	70	2000	Internal walling/ flooring units

The requirements for improvement of the properties of **cement-based materials** and **plastics** are different. In order to improve the **tensile performance** of concretes, fibres will need to

- Grip the matrix well.
- Have high E value in order to carry as much stress as possible.
- Have sufficient strength to carry imposed stresses.

Figure 2.14 shows how fibres need to carry loads across a crack. It is found that relatively small volumes of high-performance fibres can achieve these ends and in any case it is not usually easy to incorporate large quantities of fibres into cement-type materials. Table 2.3 gives typical applications. It may be surprising to see polypropylene

Fig. 2.15 *Use of steel fibre reinforced concrete to achieve stress transfer between lapped reinforcing bars. The bars in this test beam were lapped by only 7 bar diameters but the beam performed as well as one containing continuous steel.*

(plastic) fibres being used to 'reinforce' concrete since they are less stiff than the material they are reinforcing and therefore break one of the above criteria. They appear to work by arresting microcracks as they form, and this may be the critical aspect of performance in some situations. Strictly speaking, they do not 'reinforce' the material.

Figure 2.15 shows a special steel fibre-reinforced mortar being used to transfer load between reinforcing bars. The small beam shown contained 12 mm bars lapped by only 80 mm and carried a load as high as that carried by a similar continuously reinforced beam. Such arrangements permit efficient structural coupling of precast units.

As far as **plastics** are concerned, fibres are usually incorporated to **stiffen and strengthen** them. This requires in essence that fibres:

- Have high E values relative to the plastic.
- Be used in the highest volume fractions possible.

The use of a few per cent of fibres, as was the case when reinforcing cement-type materials, would be unlikely to be very effective. The following types are illustrations of fibres fulfilling these requirements:

Glass	E value, 70 kN/mm^2
Kevlar (polyamide)	E value, 100 kN/mm^2
Carbon	E value, 200 kN/mm^2

Of these, glass fibres are in most common use for reasons of cost. Some examples of glass-reinforced plastics are presented in Table 2.4.

Note that fibre volume fractions are, in general, much larger than those in the cement-based equivalents in Table 2.3. Figure 2.16 shows the roof of the Millennium

Table 2.4 Glass-reinforced plastics products: example applications

| Material | Fibre | | | Application |
	Type	Proportion by volume	Arrangement	
Polyester resin	Chopped strand mat	20%	Two dimensional	Glass-reinforced polyester (GRP); manhole mouldings claddings
Polyester resin	Long fibres	60%	Unidirectional (pultrusion)	Rod and strip for reinforcement/ prestressing applications
Polytetra fluoroethylene (PTFE)	Woven fabric	20–40%	Two dimensional	Fabric roofings

Fig. 2.16 *Millennium Dome (during construction): PTFE may contain over 30% glass fibre by volume.*

Dome, the fabric of which is formed from glass fibre-reinforced PTFE. The fabric combines good tensile properties, light transmission, resistance to dirt and excellent durability.

SELF-ASSESSMENT QUESTIONS

1. Describe the nature of the following particles and indicate how they combine to make up the atom:
   (a) proton;
   (b) neutron;
   (c) electron.
   Indicate which of these is responsible for:
   (a) nuclear stability;
   (b) chemical stability.

2. Suggest why iron is one of the more common metals in the earth's crust. Give the meaning of the term 'transition element' and explain why properties of them are quite similar.

3. Describe the bonding between the following materials:
   (a) sodium chloride;
   (b) methane;
   (c) polyethylene chains.

4. Give examples of one-dimensional, two-dimensional and three-dimensional crystals, indicating properties that ensue from these structures.

5. Describe possible ways of improving the tensile performance of cement-based materials.

3 Concrete

Chapter summary

- Manufacture, testing and properties of cements.
- Properties of aggregates for concretes.
- Proportioning of concrete mixes.
- Statistical interpretation of cube results.
- Shrinkage and durability.
- Corrosion of steel in concrete.
- Admixtures.
- Lightweight and high-strength concretes.
- Concrete surface finishes.
- Recycling concrete.

3.1 Introduction

Concrete is essentially a mixture of cement, aggregates and water. Other materials added at the mixer are referred to as 'admixtures'. Materials based on cement have the following general attractions:

- Low cost.
- Flexibility of application – for example, mortars, concretes, grouts, etc.
- Variety of finishes obtainable.
- Good compressive strength.
- Protection of embedded steel.

The following are some disadvantages/problem areas:

- Low tensile strength/brittleness.
- Rather high density (though lower density types are available).
- Susceptibility to frost/chemical deterioration (depending on type).

Environmental considerations are presented in Table 3.1.

Table 3.1 *Environmental considerations: concrete*

Consideration	Assessment
Raw material availability	Very good
Extraction	Care necessary to avoid damage to landscape
Energy used in manufacture	Cement is a high energy material
Health/safety hazards	Slight hazard from cement dust. Fresh cement is caustic – can cause burns
Recyclability	Concrete can be crushed to be reused as aggregate or for hard-core

Table 3.2 *Summary of equivalent British and European standards*

Subject	British Standard	Eurostandard
Cement specification	BS 12: *Ordinary and rapid hardening PC* BS 915: *High alumina cement* BS 1370: *Low heat PC* BS 4027: *Sulphate-resisting PC*	BS EN 197
Cement testing	BS 4550	BS EN 196
Aggregates for concrete	BS 812: *Testing aggregates* BS 882: *Specification for aggregates*	BS EN 12620
Water for concrete	BS 3148	BS EN 1008
Concrete	BS 1881: *Testing concrete* BS 5328: *Specification for concrete*	BS EN 206
Admixtures	BS 5075: *Admixtures for concrete*	BS EN 934

Cement and concrete standards

The industry is currently in the transitional stage between British and European standards. Many British Standards are still valid but will be replaced by their European counterparts during the early years of the new millennium. A summary of some of the equivalent standards is given in Table 3.2. In many cases the British Standards have been modified in anticipation of the change, but the reader should check validity to ensure that the correct specification is being referred to.

3.2 Cements

The most important are Portland cements; so named because of a similarity in appearance of concrete made with these cements to Portland stone.

Portland cements

Portland cements are hydraulic – that is, they set and harden by the action of water only.

They do not rely on atmospheric action or drying for their strength.

Manufacture

Portland cements are made by heating a finely divided mixture of clay or shale and chalk or limestone in a kiln to a high temperature – around 1500 °C, such that chemical combination occurs between them. About 5 per cent gypsum (calcium sulphate) is added to the resulting clinker in order to prevent 'flash' setting and the final stage involves grinding to a fine powder.

Composition

Although the familiar grey powder may appear to have a high degree of uniformity, it is important to appreciate that Portland cement is a complex combination of the minerals contained originally in the clay or shale and the calcium carbonate which constitutes limestone or chalk. The chief compounds are listed below, together with commonly used abbreviations.

- **Clay or shale**
  SiO_2 silica (silicon oxide), abbreviated S
  Fe_2O_3 ferrite (iron oxide), abbreviated F
  Al_2O_3 alumina (aluminium oxide), abbreviated A
- **Limestone or chalk**
  $CaCO_3$ (calcium carbonate on heating gives CaO quicklime), abbreviated C

Chemical analysis or microscopic examination of cement clinker shows that four chief compounds are present. Table 3.3 summarises the properties of these compounds, including their heat emission on hydration and typical percentages in the most commonly used cement, known as 'ordinary Portland cement' and often abbreviated OPC. The percentages do not add up to 100 because small quantities of other compounds are also present.

The properties of Portland cements vary markedly with the proportions of the four compounds, reflecting substantial differences between their individual behaviour. The proportion of clay to chalk (usually approximately 1 : 4 by weight) must be very carefully controlled during manufacture of the cement, since quite small variations in the ratio produce relatively large variations in the ratio of dicalcium silicate to tricalcium silicate. A greater proportion of chalk favours the formation of more of the latter, since it is richer in lime. This would lead to a cement with more rapid early strength development. The tricalcium aluminate and the tetracalcium aluminoferrite contribute little to the long-term strength or durability of Portland cements but would be difficult to remove and, in any case, are useful in the manufacturing process. They act as fluxes, assisting in the formation of the silicate compounds, which would otherwise require higher kiln temperatures.

Table 3.3 *Properties of the four chief compounds in Portland cements*

Name	Abbreviation	Approximate percentage in OPC	Properties	Heat of hydration (J/g)
Dicalcium silicate	C_2S	20	Slow strength gain – responsible for long-term strength	260
Tricalcium silicate	C_3S	50	Rapid strength gain – responsible for early strength (e.g. 7 days)	500
Tricalcium aluminate	C_3A	12	Quick setting (controlled by gypsum); susceptible to sulphate attack	865
Tetracalcium aluminoferrite	C_4AF	8	Little contribution to setting or strength; responsible for grey colour of OPC	420

Quite a major part of the energy cost of manufacturing cement is incurred in baking off the water contained in the raw materials. For this reason current practice (where feasible) favours the 'dry' process in which the raw materials are mixed in a dry state rather than the more traditional 'slurry' which has quite a high water content.

There will be variations in cement properties from one works to another as a result of differences in the composition of the clay or shale so that, if maximum uniformity is to be achieved in a concreting job, cement from the same works is essential.

The properties of a cement are dependent not only on its composition but also on its fineness. This is because cement grains have very low solubility in water. This could be demonstrated by mixing a small amount of cement with water in a beaker and leaving for some hours, stirring occasionally. There will be no visible change in the appearance of the cement. Cements hydrate by a surface reaction and the rate of reaction, and hence setting and hardening, therefore increase as the surface area of grains increases – that is, as they become finer. Fineness is measured by the term 'specific surface', which is defined as:

$$\frac{\text{surface area of the grains in a sample}}{\text{mass of that sample}}$$

The units of specific surface are m^2/kg, that of ordinary Portland cement (OPC) being required by earlier editions of BS 12 to be not less than 225 m^2/kg, though in practice it is usually in the range 350–380 m^2/kg. The current edition of BS 12 (1996) does not specify a minimum fineness for OPC; indeed the standard, which has been produced with the developments in European standards in mind, does not use the term 'ordinary Portland cement' though the term is still used by cement manufacturers to describe Portland cements of strength class 42.5 N as described in the standard. This strength is based on mortar cube tests.

Ordinary Portland cement is the cement best suited to general concreting purposes. It is the lowest priced cement and combines a reasonable rate of hardening with moderate heat output. Other types of cement are, however, available, each being recommended for specific applications.

Other types of Portland cement

The chief variants on ordinary Portland are given below.

Rapid-hardening Portland cement

This could be obtained by increasing the C_3S content as explained above but is normally obtained from OPC clinker by finer grinding (specific surface about 450 m^2/kg) and is covered by the same standard (BS 12) as OPC. Again whereas earlier editions of BS 12 required a minimum fineness of 325 m^2/kg, the current edition does not specify a value. Rapid-hardening Portland cement (RHPC) tends to set and harden at a faster rate than OPC. The rate of set is controlled by the addition of slightly more gypsum during manufacture, since it is the hardening properties and heat emission rather than setting rate which form the basis of applications. Early strength development is considerably higher than that of OPC, while long-term strength is similar. Applications include the following.

- To permit increased speed in construction – for example, subsequent lifts of concrete can proceed more rapidly on account of its increased early strength.
- In frosty weather there is less risk of the concrete freezing, since the cement has a higher early output.
- Concrete made with rapid-hardening cement can be safely exposed to frost sooner, since it matures more quickly.

The term 'rapid hardening Portland cement' is no longer used in the current edition of BS 12 though it is still used by manufacturers to refer to cements of strength class 52.5 N as defined in the standard.

Sulphate-resisting Portland cement (BS 4027)

This cement (SRPC) has better resistance to sulphate attack than OPC (see Sulphate attack, p. 86). The percentage of the sulphate-susceptible tricalcium aluminate (C_3A) is limited to 3.5 per cent in BS 4027 in order to minimise chemical combination with sulphates in solution. Sulphate-resisting cement may be produced by the addition of extra iron oxide before firing. This combines with alumina which would otherwise form C_3A, instead forming C_4AF which is not affected by sulphates. Hence, sulphate-resisting cement may be slightly darker in colour than OPC. Three strength classes, 32.5 N, 42.5 N and 52.5 N, are referred to in the standard. Applications include foundations in sulphate-bearing soils, in mortar for flues in which sulphur may be present from fumes and in marine structures, since sea water contains sulphates. The current standard refers to cement strength grades as in BS 12 and to a low alkali option of SRPC for use where there is a risk of alkali-silica reaction.

Low-heat Portland cement (BS 1370)
The heat output of this cement (LHPC) is limited by BS 1370 to 250 J/g at the age of 7 days and 290 J/g at the age of 28 days. These may be compared with typical heat outputs of 330 J/g and 400 J/g at the same ages for OPC. The reduction in heat output is obtained by means of lower quantities of the rapidly hydrating compounds C_3S and C_3A. In order to produce satisfactory development of strength with time, the fineness of this cement must not be less than 275 m^2/kg. Although early strength is slightly less than that of OPC, long-term strength is similar.

The chief use of LHPC is in mass concrete which might loosely be defined as concrete in which the **minimum** dimension is greater than 1000 mm. There is a tendency for the heat to build up internally in such structures, the inner core of material being much warmer than the exterior. A temperature differential of as little as 10 °C between inner and outer layers could cause internal cracking as the warmer inner layers eventually cool and are thrown into tension. Typical applications include large raft foundations and dams.

Portland blast-furnace cement (BS 146)
This cement (PBFC) comprises a mixture of OPC and ground blast-furnace slag, the proportion of the latter not exceeding 65 per cent of the total. The slag contains mainly lime, silica and alumina (in order of decreasing amounts). These exhibit hydraulic action in the presence of calcium hydroxide liberated by the Portland cement on addition of water. Early strength is lower than that of OPC but PBFC generates less heat than OPC and produces better sulphate resistance. Ultimate strength is similar to that of OPC concrete and PBFC may be competitive in cost in regions in which the slag is produced. Ground blast-furnace slag may also be used in part replacement of cement, reducing cost and modifying properties as above. Cement grades as in BS 12 are specified.

Pulverised fuel ash (PFA) in cements
PFA is a by-product of coal-powered power stations and is an example of a pozzolanic material – one which, in the presence of lime (liberated by Portland cement) has hydraulic (cementing) properties. It can be incorporated in cement during manufacture.

Microsilica
This is a by-product of electric arc furnaces used to produce silicon and ferrosilicon, most material being imported from Scandinavia. It is strongly pozzolanic and can also be incorporated with cement during manufacture. Its presence improves resistance to sulphate attack and alkali–silica reaction.

White Portland cement
This contains not more than 1 per cent of iron oxide, which is responsible for the grey colour of OPC. China clay, which is almost iron-free, is used instead of ordinary clay. Firing and grinding are also modified to prevent coloured matter being introduced. In consequence, the cement costs approximately twice as much as OPC. Setting is similar

to OPC but, being composed mainly of C_3S, strength development may be faster than that of OPC (though within the limits of BS 12). The cement is used to produce white or coloured concretes – for the latter, pigments would be incorporated. These concretes may be employed for their aesthetic or light-reflecting qualities.

Cement replacements

Three materials – blast furnace slag, PFA and microsilica – can be incorporated into concrete by adding at the mixer, each having possible cost benefits as well as contributing to improved performance, especially where there is a risk of deterioration by sulphates or other agencies. The DoE method of mix design now incorporates a section dealing with the design of such mixes.

Trends in composition/fineness of Portland cement

There is strong commercial competition between cement manufacturers who seek to supply a product resulting in maximum performance at the normal strength specification age of 28 days. In consequence there has been a trend over the last 40 years or so towards finer cements with a higher C_3S content. A concrete mix of a given specification will have a higher strength than previously since the cement is relatively fully hydrated at 28 days. Hence, to obtain a given strength, less cement will be needed, making the mix more economical. Concern has been expressed that with finer cements there will be relatively little hydration after 28 days and that subsequent strength gains will be reduced compared with 'traditional' cements. This could have possible implications for the longer term durability of the product. In consequence, BS 12 now specifies an upper strength limit for grade 42.5 cements to guard against the effects of very finely ground cements. A cement known as 'controlled fineness Portland cement' is also marketed although at present its use is mainly for some types of precast product where the coarser cement permits easier dewatering after compaction. Where there is concern about the long-term performance of the concrete, a minimum cement content could be specified.

Testing cements

A selection of tests, for setting time, soundness and compressive strength will be briefly described as they each have implications for the cement performance.

Setting time (BS EN 196–3)
In this test, a sample of cement is mixed with water to form a paste of standard consistency; checked by using a cylindrical plunger in the apparatus of Fig. 3.1. A 40 mm thick sample of this paste is then subjected to two types of penetration test, using Vicat apparatus. In the first, the penetration of a 1.13 mm diameter round needle is measured (Fig. 3.2). When this is between 4 and 6 mm from the base, the cement is said to have reached its 'initial set'. This would be related, for instance, to the time available for placing the concrete and must be not less than one hour for cements of grades 32.5 and 42.5 and not less than 45 minutes for cements of grades 52.5 and 62.5 (BS EN 197).

Fig. 3.1 *Vicat apparatus fitted with initial set needle.*

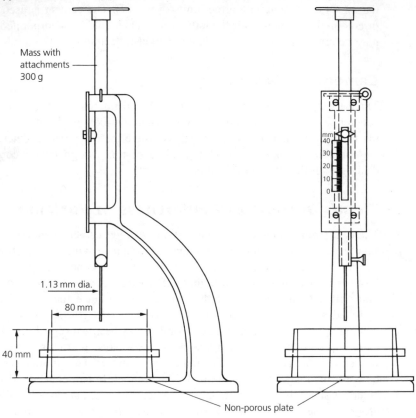

Mass with
attachments
300 g

1.13 mm dia.

80 mm

40 mm

Non-porous plate

Fig. 3.2 *Enlarged view of the final set needle for Vicat test.*

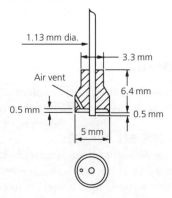

1.13 mm dia.

3.3 mm

Air vent

6.4 mm

0.5 mm

0.5 mm

5 mm

At a much later stage, a 1.13 mm diameter round needle projecting 0.5 mm from a small brass cylinder is applied (Fig. 3.2) and, when the brass cylinder fails to mark the cement paste, the 'final set' has occurred. This would be related to the time after which concrete could be treated as 'solid', though BS EN 197 no longer gives requirements for

Fig. 3.3 *Le Châtelier apparatus for measuring soundness of cement.*

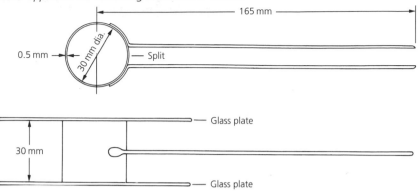

the final setting time. It should be emphasised that both 'initial' and 'final' sets are arbitrarily selected points on a smooth and continuous curve relating stiffness to time.

Soundness (BS EN 196–3)

This test is primarily designed to detect the presence of any free lime which might be present inside grains of cement clinker. Such lime would, if present, gradually become exposed on hydration of the cement. It then 'slakes' due to reaction with water, resulting in an overall expansion (unsoundness) in the hardened concrete with a consequent risk of damage. Since the exposure of the lime takes some months to occur, an accelerated test is used to detect its presence. The Le Châtelier apparatus (Fig. 3.3) is used, consisting of a small brass cylinder containing a split, on each side of which are fixed a long pointer to magnify any movement. A sample of cement mortar is allowed to harden in the apparatus and then boiled for three hours. The expansion is noted after cooling and should not be greater than 10 mm. If the cement fails to comply, a sample is aerated by exposure to air of humidity between 50 and 80 per cent for 7 days. This causes slaking of any lime, to which air has ready access. The expansion of the cement on repeating the soundness test should not exceed 10 mm.

Compressive strength (BS EN 196–1)

The current test for compressive strength of cements relates to samples of size $40 \times 40 \times 160$ mm, made from a 3 : 1 sand/cement mortar with a water/cement ratio of 0.5. The sample is first tested in flexure and then the broken pieces are tested in compression using a special 40×40 mm jig. The test employs a standard CEN sand of specified grading, the particle sizes being between 1.6 mm and 80 µm. A sand/cement ratio of 3 : 1 is employed with a water/cement ratio of 0.4. Standard sands such as Leighton Buzzard sand should be checked against the EN 196 specification before being used. The mortar is prepared by mechanical mixing. Requirements for the compressive strength of mortar cubes are given in Table 3.4.

Note that an extra letter is added after the strength class to denote the rate of early strength development. Most common cements are given the letter 'N' corresponding to normal development, while the letters 'L' and 'R' are used to denote low and rapid rates of development respectively.

Table 3.4 *Strength requirements of mortar cubes*

Data taken from BS 12 (1996)

Class	Early strength (N/mm²)		Standard strength (N/mm²)	
	2 days	7 days	28 days	
32.5N	–	≥ 16	≥ 32.5	≤ 52.5
42.5N	≥ 10	–	≥ 42.5	≤ 62.5
52.5N	≥ 20	–	≥ 52.5	≤ 72.5

3.3 Structure of hydrated cement

The most important components of Portland cement from the strength development point of view are C_2S and C_3S which, on hydration, form the same compounds in differing proportions:

$$2(2CaOSiO_2) + 4H_2O \longrightarrow 3CaO \cdot 2SiO_2 \cdot 3H_2O + Ca(OH)_2$$

calcium silicate calcium silicate hydrate calcium hydroxide

$$2(3CaOSiO_2) + 4H_2O \longrightarrow 3CaO \cdot 2SiO_2 \cdot 3H_2O + 3Ca(OH)_2$$

calcium silicate calcium silicate hydrate calcium hydroxide

(Note that the full formulae for C_2S and C_3S have been given here to enable the equations to be balanced.)

The physical form of these compounds plays a most important part in the behaviour of concrete. The calcium silicate hydrate forms extremely small fibrous, platey or tubular crystals, which can be regarded as a sort of rigid sponge, referred to as cement 'gel'. This gel must be saturated with water if hydration is to continue. The calcium hydroxide forms much larger platey crystals. These partially dissolve in water in damp conditions, providing hydroxyl (OH⁻) ions, which are important for the protection of steel in concrete. As hydration proceeds, the two crystal types become more heavily interlocked, increasing the strength of the concrete, though the main cementing action is provided by the gel which occupies two-thirds of the total mass of the hydrate.

3.4 High-alumina cement

It is important to appreciate that high-alumina cement (HAC) is not a Portland cement and has quite different properties. It was developed in the 1930s to resist sulphates and is manufactured from limestone and bauxite (aluminium ore). The main cementing ingredient is mono-calcium aluminate (CA using cement notation) and since C_3A is not present, the cement normally has good sulphate resistance. It also exhibits very rapid hardening properties (but is not quick setting). In consequence, the cement has been widely used for precasting, since the turn-round time of expensive steel moulds is reduced. The design of HAC concrete is in fact based on 24-hour strength rather than

28-day strength for OPC concrete. Concretes made with HAC are usually a darker colour than Portland cement equivalents, reflecting the dark grey colour of the cement. Use of HAC concrete for structural purposes in building has been banned since 1973 following the collapse of a school roof and swimming pool roof. At least partly responsible for the collapse was the phenomenon of 'conversion'. This is the name given to a rearrangement of the crystalline hydrate which results in the formation of pores in the concrete. This leads in turn to permeability and reduced strength. Conversion was initially thought to be associated with high temperatures and humidities (as might occur in a swimming pool environment), though it is now known to occur slowly in normal conditions. The strength reduction can be minimised by use of low water/cement ratio concretes. These concretes not only have higher initial strength, but on conversion it is possible that residual unhydrated cement can hydrate, filling pores as they are formed. There are many HAC structures still in existence, but these should be periodically checked to determine the degree of conversion and the consequent effect on strength. On heating to high temperatures HAC forms a ceramic bond and is hence widely used for refractory purposes such as flue linings.

3.5 Aggregates for concrete

Aggregates are used in concrete for the following reasons:

- They greatly reduce cost.
- They reduce the heat output per unit volume of concrete and therefore reduce thermal stress.
- They reduce the shrinkage of the concrete.
- They help to produce a concrete with satisfactory plastic properties.

There may be other reasons for which 'special' aggregates would be employed; for example:

- Low-density concrete (2000 kg/m^3 or less) to decrease foundation loads, increase thermal insulation and reduce thermal inertia (lightweight aggregates).
- High-density concrete (2600 kg/m^3 or more) as required, for example, for radiation shielding (barytes – barium sulphate – or iron-based aggregates).
- Abrasion-resistant concretes for floors (granite or carborundum aggregates).
- Improved fire resistance (limestone, lightweight aggregates such as expanded pulverised fuel ash).
- Decorative aggregates – for example, crushed granite is available in several different colours which can be revealed by use of an exposed aggregate finish.

Maximum aggregate size

As a general rule

> The maximum size of an aggregate should be as large as possible, consistent with minimum section size and reinforcing bar spacings.

Table 3.5 *Concrete aggregate sizes: typical applications*

Nominal max. size (mm)	Application
40	Mass concrete, road construction
20	General concrete work, including reinforced and prestressed concrete
10	Thin sections, screeds over 50 mm thickness
5	Screeds of 50 mm thickness or less

Larger aggregates result in a lower sand requirement. This reduces the specific surface of the aggregate overall and consequently reduces water and cement requirements. The resulting concrete should therefore be more economical and exhibit lower shrinkage for a given strength.

There are, of course, constraints on the maximum size that can be employed; first, the maximum aggregate size should not exceed approximately one-fifth of the minimum dimension in the structure, consideration also being given to spacings between reinforcing bars. Secondly, the aggregate size chosen will need to be a size that is readily available. The standard sizes, together with typical applications, are given in Table 3.5.

Aggregates which are largely retained on a 5 mm mesh sieve are described as coarse aggregates. These may comprise uncrushed natural gravels which result from the natural disintegration of rock or crushed stones and crushed gravel, produced by crushing hard stone and gravel respectively. Aggregates that largely pass a 5 mm sieve are referred to as **sands**. (*The current edition of BS 882, 1992, uses the term 'sand' in preference to 'fine aggregate' though the latter term is retained in the latest edition of the DoE method of mix design. The term 'fines' is reserved for material such as silt or clay, passing the 75 μm sieve.*) Sands may comprise natural sands; that is, sands resulting from natural disintegration of rock or crushed stone, or crushed gravel sands.

Grading of aggregates

Aggregates for concrete are usually graded continuously from their maximum size down to the size of the cement grains, since this ensures that all voids between larger particles are filled without an excess of fine material. (It will be recalled that an excess of fine material results in increased water and cement requirements, as explained above.) This effect is illustrated in Fig. 3.4.

Both crushed and uncrushed natural gravels have naturally continuous gradings, but to comply with BS 882, the gradings must be within certain limits. Gradings are determined by passing aggregates through a set of standard sieves, the particular sieves used depending on the maximum aggregate size (see Experiment 3.1). The grading is then defined by the percentage of the total sample used which passes each sieve. Table 3.6 shows the sieves used for concreting sands, together with specimen results. These are

Fig. 3.4 *Simple illustration of the use of graded aggregate in concrete. Spaces between larger particles are filled with progressively smaller particles.*

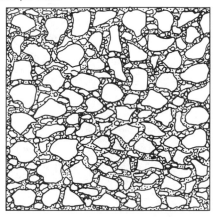

Table 3.6 *Example sieve analysis of a sand used for concreting*

Sieve size	Mass retained (g)	Cumulative mass retained (g)	Mass passing (g)	Percentage passing
10 mm	0	0	287	100
5	6	6	281	98
2.36	17	23	264	92
1.18	32	55	232	81
600 μm	48	103	184	64
300	81	184	103	36
150	86	270	17	6
Passing 150	17	287	–	–

plotted in Fig. 3.5; note that the sieve sizes are not plotted to a linear scale, successive sizes being spaced equally instead, and points are joined by straight lines. Figure 3.5 also includes the limits of BS 882 grading F into which this particular sand fits. The current edition of the standard gives three gradings, C, M and F, corresponding to fine, medium and coarse sands, replacing the four 'zones' of earlier editions. This reflects the fact that the previous zones were rather restrictive in terms of sand acceptance; materials which did not comply with a 'zone' often being perfectly acceptable for use in concrete, with suitable mix proportioning. There are no rigid distinctions between the C, M and F gradings and an aggregate may comply with two of the envelopes. The aggregate shown in Fig. 3.5, for example, complies with grading M as well as with grading F. However, very fine gradings may cause problems and BS 882 states that sands should comply with gradings C or M if being used for heavy duty concrete floor finishes.

Fig. 3.5 *Sand grading corresponding to sieve analysis results given in Table 3.6. The F grading envelope into which this sand fits is also indicated.*

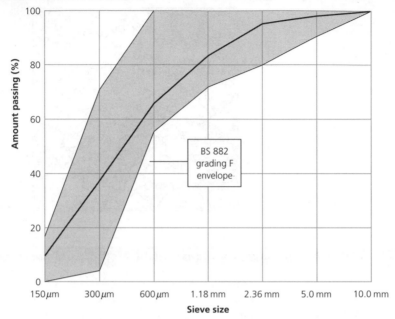

Coarse aggregates are not assigned grading 'zones', since the range of particle size is, in general, smaller than that of sands. They should, however, fit into a BS grading envelope for that particular size and type – see Experiment 3.1.

Table 3.7 indicates that there are three ways in which a continuously graded aggregate of maximum size 20 mm may be stockpiled and batched. The simplest method would be to have a single stockpile of 20 mm 'all in' aggregate, sometimes referred to as 'ballast', since this material is relatively cheap and, apart from the cement, only a single quantity then needs to be batched. This method gives, however, only poor control over the distribution of particle sizes within each batch quantity. Coarse particles are

Table 3.7 *Batching graded aggregate: three alternatives*

Maximum size 20 mm

	Aggregate size			
No. of stockpiles	*20 mm*	*10 mm*	*5 mm*	*150 μm*
1	20 mm 'all in' aggregate (ballast)			
2	20 mm graded coarse aggregate			sand
3	20 mm single-size coarse aggregate	10 mm single-size coarse aggregate		sand

prone to separation from fine material, especially if stockpiling technique is poor or if aggregates dry out while stockpiled. Successive batches of the resulting concrete would therefore tend to have variable water requirements for a given workability, depending on the fineness of individual batches of aggregate. While this method of batching is satisfactory for nominal lightly stressed concrete – such as foundations and ground floor slabs in domestic dwellings – it would not be recommended for structural quality concrete. The second alternative is to use '20 mm graded coarse aggregate' and sand, thereby ensuring that the ratio of quantities of aggregate of sizes above and below 5 mm is maintained accurately constant. This 'two-stockpile' method is widely used in concrete production.

Even closer control results when three stockpiles are used:

- 20 mm single-sized coarse aggregate.
- 10 mm single-sized coarse aggregate.
- Sand.

The subdivision of coarser sizes which are most prone to segregate permits very close control over the material as a whole. The extra cost of purchasing a higher quality material and additional batching may, in structural applications, be offset by the reduced variability of the resultant concrete.

Aggregate density

Before classifying aggregates into density groups, it is necessary first to examine the meaning of the term 'density', since a number of differing definitions exist, according to the context of use.

Relative and solid density

Most aggregates comprise a mass of solid material containing air pores which may or may not be accessible to water. The term which describes the density of the solid material itself is **relative density** (formerly called specific gravity) or density relative to that of water, though several definitions are possible according to the way in which accessible pores are treated.

Perhaps the most important of these is relative density in the **saturated surface dry** (SSD) state – that is, when all accessible pores are full of water but the aggregate surface is dry. This relative density usually forms the basis of mix design methods and is defined as:

$$\text{relative density (SSD)} = \frac{\text{mass of given sample of SSD aggregate}}{\left(\begin{array}{c} \text{volume of water} \\ \text{displaced by saturated} \\ \text{surface dry sample} \end{array} \right) \times \text{density of water}}$$

Note that because it is a ratio, relative density has no units.

Relative density could be defined in other ways – for example, by taking the mass of dry aggregate with the same volume as given above. (This is referred to as the **oven dry** relative density and gives a lower result because the mass in the numerator of the equation does not include the mass of water as previously.)

In the case of most natural aggregates, void contents are small, so that differences between the various definitions of relative density are correspondingly small. The term 'solid density' will be taken to mean 1000 × relative density (say, on a saturated surface dry basis), normally having the units kg/m³.

Bulking and bulk density

When aggregates are loosely packed together or stockpiled, large volumes of air are trapped between particles – usually many times the volume of air present within particles. This is referred to as 'bulking' and for coarse aggregates it amounts to between 30 and 50 per cent of the total space occupied. The extent of bulking of sands depends very much on their moisture content. The void content of dry sand may be quite small – say, 20 per cent since the particles of different sizes exhibit a void filling action. This can increase to 40 per cent when 5–10 per cent moisture is present, thereafter decreasing as further moisture tends to assist particles to consolidate. The bulking at intermediate moisture contents is the result of thin water films increasing friction between sand particles. The situation is illustrated in Fig. 3.6 (see Experiment 3.2).

Fig. 3.6 *Volume occupied by a given mass of sand when (a) dry, (b) damp and (c) wet.*

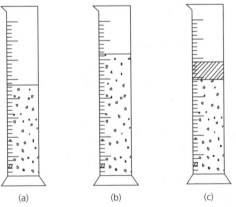

(a)   (b)   (c)

Bulking of aggregates produces uncertainty in the solid content of aggregates batched by volume and for this reason batching by weight (mass) is much preferred, except for concrete having only nominal performance requirements. Hence, most concrete production equipment is geared to weight batching.

Low-, medium- and high-density aggregates

Most natural aggregates have solid densities within quite a narrow range of values, between 2400 and 2700 kg/m³ (corresponding to relative densities between 2.4 and 2.7). These would result in concrete having densities normally in the range 2200–2500 kg/m³, slightly lower than the corresponding aggregate densities on account of the water content of the concrete.

Lightweight aggregates comprise highly porous particles and are widely used in the construction industry. They have the advantages that:

- Less raw material is used in producing the concrete.
- Foundation loads are reduced.
- Thermal properties are improved.

Examples of synthetic lightweight aggregates are sintered pulverised fuel ash, expanded clay, blast-furnace slag and expanded slate. Natural lightweight aggregates such as pumice may also be used. Densities as low as 500 kg/m^3 can be obtained in this way (see 'Lightweight concretes', p. 93).

There are certain situations in which high-density concretes are required – for example, for radiation shielding for nuclear reactor vessels. Such concretes are produced by use of high-density aggregates such as:

- **Barytes** (barium sulphate); relative density 4.1.
- **Magnetite** (iron ore) relative density 4.5.
- **Metallic** aggregates such as steel shot relative densities over 7.0.

Metallic aggregates may result in concrete having over twice its normal density.

Silt and clay (defined in BS 882 as 'fines')

These may be defined as material passing a 75 μm sieve (compared with the smallest BS sieve size for sand, 150 μm).

> Silt and clay are harmful to concrete if present in substantial amounts, since they increase the specific surface and hence the water requirement of a mix, resulting in lower strength unless the cement content is also increased.

Sands are likely to contain more silt than coarse aggregates and a rapid estimate of the silt/clay content of sands (other than crushed stone sands) can be made by a 'field settling test'. The test involves shaking a measuring cylinder containing a 1 per cent salt and sand mixture (see Experiment 3.3). The salt helps to separate the silt into a separate layer whose volume can be measured after being allowed to stand for three hours (Fig. 3.7). If the amount is more than approximately 8 per cent by volume, separate tests are necessary to determine whether the material passing the 75 μm sieve by weight is in excess of the BS 882 limit of 4 per cent for natural sand and crushed gravel sand or 16 per cent for crushed rock sand. Coarse aggregates can be checked for silt and clay by visual inspection. When in doubt, the test of BS 812: Part 103.1 should be used to ascertain whether the silt content is in excess of BS 882 limits.

Moisture contents

It is advisable during quality concreting to check moisture contents of aggregates prior to batching, so that accurate adjustment of batch weights can be made where necessary. The amount of free-water in aggregates should be deducted from the water to be added, and batch quantities of aggregates must be slightly increased if free water is present since some of the 'aggregate' is actually water. (A further possibility, though this is not common with natural aggregates in the UK, is that extra water may be required if

Fig. 3.7 *Field settling test for detection of silt/clay in sand.*

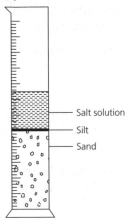

— Salt solution
— Silt
— Sand

the aggregates are dry, such that they absorb water on wetting. The latter situation is quite common with lightweight aggregates which absorb substantial quantities of water.) Sands normally trap more water than is trapped by coarse aggregates, since there are more points of contact between particles.

Aggregate stockpiles are normally wet, assuming that they are replenished regularly, though the surface may dry, especially in warm windy weather. In this situation there may be quite large variations in moisture content from place to place, requiring additional care when batching.

The principles of three types of test are given below and in each case the aim will be to obtain the free-water content of the aggregates since no adjustment is necessary in mixing water for moisture which is absorbed within aggregate particles.

Drying methods
In these a sample of damp aggregate is carefully weighed (M_w) and dried until it reaches a saturated surface dry condition. A suitable method is by hairdryer, sand being dried uniformly until it is just free flowing. Coarse aggregate should be dried until all traces of surface moisture have disappeared. A frying pan on a gas or electric ring may alternatively be used (though care is necessary to ensure uniform heating of the sample). The aggregate is then weighed again (M_d). The moisture based on wet weight is then given by:

$$\text{free-moisture content} = \frac{M_w - M_d}{M_w} \times 100 \quad \text{(per cent)}$$

Siphon can
This (see Experiment 3.4) relies on the fact that the volume of an accurately measured mass of aggregate of given density depends on the amount of moisture in it, water having a density of only approximately 40 per cent that of most natural aggregates. Hence the greater the moisture content, the greater the volume of a given mass (normally 2 kg). A sample of dried aggregate is necessary for the test. To give the **free-water content** this should be in the **saturated surface dry** condition. Accurate weighing of

dry and damp samples is necessary in this method, which is suitable for both coarse aggregate and sand.

Calcium carbide method

A rapid estimate of the moisture content of sand can be made by mixing a standard quantity of aggregate with an excess of calcium carbide in a pressure vessel (see Experiment 3.5). The calcium carbide reacts with moisture, releasing a gas whose pressure is measured. The pressure scale is calibrated to read moisture content based on wet weight directly. This method measures the free-moisture contents of hard materials such as aggregate, since the calcium carbide cannot easily penetrate pores in the stones. The sample size for this test is quite small so that careful sampling is necessary for sand and the test is not suitable for coarse aggregate – there would be too few particles to give a representative sample.

Correction for moisture content

A method of correction of batch quantities for a given volume of concrete will be illustrated by means of a worked example. Determine the correct batch quantities for 1 cubic metre of concrete if the quantities based on saturated surface dry aggregates are:

Cement	310 kg
Sand	650 kg
Coarse aggregate	1190 kg
water	180 litres

Moisture contents of sand and coarse aggregates are 4.5 and 1.5 per cent respectively, based on wet weight.

Answer

The moisture in 650 kg of sand equals

$$\frac{4.5}{100} \times 650 = 29.25 \text{ kg}$$

Hence, add this amount to the quantity of sand giving

650 + 29.25 = 679.25 (say 680) kg

as the correct batch mass of sand. The moisture in 1190 kg of coarse aggregate equals

$$\frac{1.5}{100} \times 1190 = 17.85 \text{ kg}$$

Again this amount must be added to the quantity of coarse aggregate, giving

1190 + 17.85 = 1207.85 (say 1208) kg

as the correct batch quantity of coarse aggregate. To obtain the correct batch quantity of water, the water contents above are subtracted from the given batch quantity:

180 − (29.25 + 17.85) = 132.9 (say 133) kg, or litres

It may be argued that the above method is inaccurate, since the extra material added to sand and coarse aggregates also contains water, which should be allowed for. Strictly speaking this is true, but the errors in percentage terms are small enough to be ignored and the actual errors would normally be smaller than the tolerances in batching equipment. The above method is given on account of its simplicity.

3.6 Water for concrete

Water is the chemical means by which cement is converted from a powder into a hardened material with strength and durability. It therefore needs to be of appropriate quality and two main aspects may need to be considered.

Organic contamination

Water which is in contact with organic matter such as vegetation can be contaminated by organic acids. These can reduce the rate of hydration by increasing the pH value (acidity) of the wet concrete. Water which appears green due to the presence of algae can lead to air entrainment which reduces strength.

Dissolved salts

Sea water, for example, may contain dissolved salts such as sulphates which can react with the hydrated cement, or chlorides which tend to accelerate hydration and increase the risk of corrosion of embedded steel. Where chlorides are present the salt content of the water should be added to that of the aggregates to obtain a combined total figure. This should then be checked against allowable values.

In general, clear, flowing water or drinking water should be suitable for production of concrete but, if in doubt, cubes can be made to the required specification and checked against similar cubes made with distilled water. BS 3148 gives details.

3.7 Principles of proportioning concrete mixes

Consideration has already been given to properties of cements and aggregates and it is now necessary to identify the important properties of the concrete itself in the fresh and hardened states and to examine the effect of materials types and proportions on them.

Perhaps the most important single term in understanding how mix parameters affect concrete properties is the water/cement ratio.

$$water/cement\ ratio = \frac{mass\ of\ water\ in\ a\ concrete\ sample}{mass\ of\ cement\ in\ the\ sample}$$

Note that the **free**-water/cement ratio is the most important ratio since it is this that affects strength and durability. Free-water/cement ratio would be based on the free-water content of the mix – that is, water absorbed in the aggregates is disregarded.

A further term traditionally used is the aggregate/cement ratio.

$$aggregate/cement\ ratio = \frac{mass\ of\ aggregate\ in\ concrete\ sample}{mass\ of\ cement\ in\ that\ sample}$$

For most concrete mixes the aggregate/cement ratio would be in the range 4–10, small ratios indicating rich, expensive mixes and high ratios lean, cheaper mixes. Although mortars are still commonly specified in terms of aggregate (sand)/cement ratios the richness of concrete mixes is now more commonly specified in terms of cement content per cubic metre. For example, an aggregate/cement ratio of 4 would correspond to about 450 kg of cement per cubic metre while an aggregate/cement ratio of 10 would correspond to about 200 kg of cement per cubic metre. One advantage of expressing the richness of mixes in terms of cement content is that the cost of the concretes of varying richness can be relative easily compared, cement being the most costly item.

Properties of fresh concrete

The fresh concrete must be satisfactory in relation to its workability and its cohesion

Workability
This the most important term relating to fresh (plastic) concrete.

Workability is that property of the concrete which determines its ability to be placed, compacted and finished.

Of these three operations, the greatest emphasis should be placed on compaction (elimination of air voids), since the consequences of inadequate compaction are serious.

(*Note that the new BS EN 206 uses the term consistence instead of workability, though the latter term will be retained here in view of its current continued use in the industry.*)

Workability may be measured by the slump test, compacting factor test and Vebe test (BS 1881: Parts 102–104), for which brief details of each will be given. For each test it is essential that a representative sample of concrete be obtained. Details of how to obtain such samples are given in BS 1881: Part 101.

In the slump test (see Experiment 3.6) concrete is placed into a special cone in three layers of equal height, each layer being compacted in a standard manner. On removing the cone the concrete 'slumps', the slump being equal to the difference between the height of the cone and the highest point on the concrete (Fig. 3.8). When the concrete slumps evenly, a 'true' slump is said to occur. In less cohesive concretes, collapse may occur on an inclined shear plane, producing a 'shear' slump. A wet mix of low cohesion may produce a 'collapse' slump. Slumps may vary from zero for dry concrete mixes, through 50–70 mm for medium workability concretes to 150 mm or more for wet or harsh mixes. The test is widely used as an on-site check, though it is not suitable for dry mixes which tend to give low or zero slump.

Fig. 3.8 *Method of measuring slump in slump test together with types of slump: (a) true slump; (b) shear slump; (c) collapse slump.*

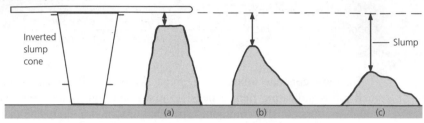

Fig. 3.9 *Compacting factor test.*

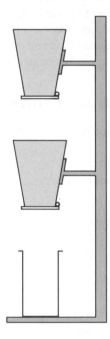

In the compacting factor test (see Experiment 3.7) concrete is first gently loaded into a hopper (Fig. 3.9), is then allowed to fall vertically into a similar lower hopper and finally into a cylinder, by the action of gravity. The concrete in the cylinder is then struck off level and weighed. The cylinder is emptied and refilled, this time fully compacting the concrete. The ratio

$$\frac{\text{mass of partially compacted concrete}}{\text{mass of fully compacted concrete}}$$

is known as the 'compacting factor'. The value will approach unity for wet mixes (for which the test is not suitable) and may be as low as 0.65 for very dry mixes.

The third test, the Vebe test (see Experiment 3.8) utilises a slump cone in a cylindrical container fixed to a small vibrating table (Fig. 3.10). A slump test is first carried out and a Perspex disc attached to a vertical guide is allowed to rest gently on the

Fig. 3.10 *Vebe consistometer.*

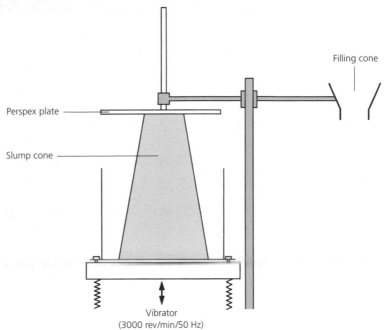

concrete. The table is then made to vibrate and the time taken for the underside of the Perspex disc to become completely covered in concrete is measured in seconds. The time may vary from a few seconds for wet mixes (for which the test is not suitable) to over 20 seconds for dry concretes.

Comments on workability tests

From the descriptions given above it will be evident that the slump test is the most suitable for the measurement of medium to high workabilities. The test has the advantage that the equipment is simple, portable and does not require an electricity supply.

The compacting factor test is sensitive to medium and low workabilities, though its use is not included in the current DoE design document *Design of normal concrete mixes* since 'it is not possible to establish consistent relationships between it and the slump or Vebe tests'. The compacting factor test may nevertheless be useful as a control test for drier concretes and is used in civil engineering applications.

The Vebe test has the disadvantage of requiring mains electricity, though it is very sensitive to changes in the water content of dry mixes, this providing the main basis for its use.

It is important to appreciate that all the above tests are used to check the workability of the concrete, once mixed; and that, since design methods rarely give a very accurate water batch quantity, even with moisture content correction, there is a need to be able to judge the workability of the concrete while in the mixer. The mixer operator may well be able to do this, for example, by the appearance of the concrete, the sound made by

Table 3.8 *Workability categories: properties and applications*

Workability category	Slump (mm)	Compacting factor	Vebe time (seconds)	Applications
Extremely low	0	0.65–0.7	Over 20	Lean mix concrete for roads (compacted by vibrating roller); precast paving slabs
Very low	0–10	0.7–0.75	12–20	Roads compacted by power operated machines
Low	10–30	0.75–0.85	6–12	High-quality structural concrete; mass concrete compacted by vibration
Medium	30–60	0.85–0.95	3–6	Normal purposes – reinforced concrete compacted by vibrating poker or manually
High	60–180	0.95–1.0	0–3	Areas with congested reinforcement, concrete for placing under water

the concrete within the drum, or the sound of the engine or motor which is used to drive the mixer, since drier mixes may increase the power consumption. Hence an experienced mixer operator is a very useful part of the construction team in producing concrete of consistently correct workability.

Workability requirements

The workability of concrete for a given situation should be the minimum value that will ensure full compaction with the plant available, due consideration being given to difficulties that may arise from access, congested reinforcement or depth of pour. Table 3.8 gives the main workability categories and applications.

Factors affecting workability

The chief factor affecting workability is water content (usually expressed as volume in litres of water per cubic metre of concrete)

For a given aggregate type and size, workability is highly sensitive to changes of water content – for example, in the case of 20 mm uncrushed aggregate, a water content of 160 litres per cubic metre would result in low workability, while a water content of 200 litres would result in high workability.

The following factors also affect workability:

- **Maximum aggregate size**
  A smaller maximum aggregate results in a higher specific surface (surface area per unit mass) of the mix overall. Therefore more water is required to 'wet' the larger surface areas involved. For a given aggregate type, decreasing the maximum size from 20 to 10 mm would increase the water requirement by approximately 10 per cent. Increasing the maximum aggregate size from 20 to 40 mm would clearly have the reverse effect.
- **Aggregate shape**
  Crushed (angular) aggregates similarly result in a larger specific surface than uncrushed (irregular) aggregates and therefore an increase of 15–20 per cent in water requirement.
- **Aggregate surface texture**
  Aggregates having a coarse surface texture will also increase water requirements, though it should perhaps be added that, for high-strength concretes, angular shapes combined with a rough surface texture allow fuller exploitation of a hard aggregate. Conversely, the strength of concretes made with natural flint-based gravels is limited at the high-strength end by a tendency of smooth aggregate surfaces to debond from the cement mortar under stress.

The workability of concrete can also be greatly influenced by incorporating **plasticisers.** Their use in the industry is therefore increasing. Plasticisers effectively make the water 'wetter' so that less is required. A 10 per cent reduction is easily achievable at constant workability. (The action is similar to that by which the cleansing power of water is improved by the addition of soap, which reduces the surface tension of water). Even greater effectiveness can be achieved by use of superplasticisers which are used at higher dosages and achieve 20 per cent or greater reductions in water without loss of workability. These materials can be used to enhance the quality of concrete provided they are used correctly and not used as an economy measure to reduce the amount of cement in the concrete (see Admixtures, p. 90).

Cohesion

Though perhaps not of such critical importance as workability, the choice of concrete of correct cohesion is an essential part of the mix design process. Cohesion (or cohesiveness) may be defined as the ability of the fresh concrete to resist segregation, though a cohesive mix will also be easier to finish than a relatively harsh mix. The cohesion in a concrete mix is provided by the fine material – that is, the finer fractions of sand and the cement. It follows that richer mixes which have a higher cement content should, in general, have a reduced sand requirement and this is obtained when concrete is designed by the DoE method of mix design. It should perhaps also be stated that cohesion should be the minimum required for the purpose. If, in common with workability, cohesion is higher than necessary, the water content of the mix will also be unnecessarily high, leading to the disadvantages described earlier.

Properties of hardened concrete

In the context of mix design the most important properties of concrete are **strength** and **durability.**

Strength

This is normally considered to be the most important property in relation to mature concrete. In the UK, 'strength' most commonly means compressive strength as measured by cubes manufactured, cured and tested according to BS 1881 (see Experiment 3.9).

(Note that the European Standard (BS EN 206) will refer to cylinders also, which in general give lower strength results for a given concrete mix than cube test results. When specifying strength it is therefore important that the type of test required is included in the description. All references made to strength below will be to compressive cube strength.)

The strength of concrete is affected by the following aspects of mix materials and proportions:

- **Free-water/cement ratio**
  As the quantity of free-water in a mix increases in relation to the quantity of cement, the density and, consequently, the strength of the concrete decrease. At very low water/cement ratios (e.g. below 0.4) a significant part of the cement never hydrates, due to inadequate space and/or inadequate water. Since unhydrated cement is of high strength, the resulting product will also be of high strength, provided full compaction is achieved. At water/cement ratios above about 0.4, the expansion of cement on hydration is insufficient to occupy the space previously filled with water. Hence porosity increases and strength decreases. At water/cement ratios in the range 0.3–1.0, strength increases progressively with decreasing water/cement ratio. At values below 0.3 the strength of fully compacted concrete continues to rise, reaching values over 100 N/mm^2 at 0.2 or less, though the concrete is very difficult to compact at such values. Plasticisers or superplasticisers would be used to help achieve these high-strength levels.

- **Aggregate properties**
  Crushed aggregates generally result in higher strength than uncrushed aggregates, since they form a better key with hydrated cement. To obtain very high strengths the use of crushed aggregates may be essential.

- **Cement type**
  Where high early strength is required, the use of rapid-hardening Portland cement (52.5 N) may be considered, though long-term strength will be similar to that of OPC (42.5 N).

Durability

The durability of normal concrete depends mainly on its permeability, and hence on the porosity of the hydrated cement.

The arguments determining durability are similar to those given under strength, low water/cement ratios resulting in greatest durability. To illustrate this, concretes of free-water/cement ratio 0.4, 0.6 and 0.8 would result in very good, moderate and poor durability respectively in concrete subject to average exposure. It is not always easy to measure the free-water/cement ratio in production, so that durability is sometimes specified alternatively by means of minimum cement content on the basis that, for a given workability, the water content would be constant; hence, by specifying a cement content, one is effectively specifying a water/cement ratio. Another way of specifying durability indirectly is by specifying strength, though the correlation depends on aggregate properties.

3.8 Prescriptive concrete mixes

There are various ways of proportioning concrete mixes, but they all converge to the same point – that is, the identification of batch quantities for a given volume of concrete. The simplest way is by **prescription**, in which mixes are specified by referring directly to their constituent quantities.

Nominal mixes

This is the simplest and least rigorous way of specifying concrete. For oversite concrete or strip foundations it may be sufficient to specify ratios such as

> 1 part cement, 2 parts sand, 4 parts coarse aggregate

or, even simpler

> 1 part cement, 6 parts all-in aggregate (ballast)

The aggregate size and workability should also be stated. It may be satisfactory to batch concrete by volume (for example, using gauging boxes) for such mixes. They may produce perfectly adequate concrete, but offer little control over variations and so would not be acceptable where structural concrete is specified.

Standard mixes

Examples of these can be found in BS 5328. Table 3.9 gives batch quantities for one cubic metre of concrete for 20 mm aggregate. The quantities shown will 'normally' give characteristic strengths stated in the table. Volume batching is permitted for mixes ST1, ST2 and ST3. This procedure permits a simple approach to specifying structural concrete, albeit of limited strength–maximum characteristic value 25 N/mm². Note the simple numerical differences between the figures in the table. Cube testing is not required but might be specified as a quality control tool.

Designated mixes

These are recommended mixes for specific applications. BS 5328, for example, defines a 'GEN 1' concrete suitable for a range of applications including

- Blinding and mass concrete fill
- Strip footings
- Mass concrete foundations

Data taken from BS 5328

Standard mix	28 day characteristic strength (N/mm²)	Constituents	75 mm slump	125 mm slump
ST1	7.5	Cement (kg)	210	230
		Total aggregate (kg)	1940	1880
ST2	10.0	Cement (kg)	240	260
		Total aggregate (kg)	1920	1860
ST3	15.0	Cement (kg)	270	300
		Total aggregate (kg)	1890	1820
ST4	20.0	Cement (kg)	300	330
		Total aggregate (kg)	1860	1800
ST5	25.0	Cement (kg)	340	370
		Total aggregate (kg)	1830	1770
ST1 ST2 ST3		Sand (percentage by mass of total aggregate)	35–50	35–50
ST4 ST5		Sand (percentage by mass of total aggregate):		
		sand grading C	35–45	35–45
		sand grading M	30–40	30–40
		sand grading F	25–35	25–35

- Trench fill foundations
- Unreinforced house floors (under screed)
- Drainage applications

In all the above applications standard mix ST2 is identified with workability appropriate to application. Hence designated mixes provide a simple way of specifying concrete for common applications.

3.9 Designed concrete mixes

Designed mixes offer the possibility of deriving mix proportions which are 'just right for the job'. They involve extra work in design and testing but, for large volume applications or high-strength concrete, produce concrete of consistently suitable properties.

Fig. 3.11 *Simple flow chart for the design of concrete mixes.*

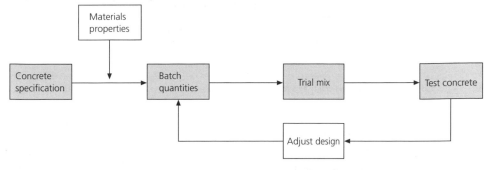

The object of mix design is, by systematic analysis of materials properties and knowledge of how they affect concrete properties, to produce, as economically as possible, batch quantities for a given volume of concrete such that the properties of both the fresh and hardened material are as required for the specific purpose.

The process can be summarised as shown in Fig. 3.11. It must be emphasised, however, that the process is iterative since mix design methods cannot allow for small variations in properties of the materials or the techniques employed.

3.10 Department of the Environment method of mix design (1988 edition)

(Note that a new edition of the method was produced in 1997, mainly to reflect the changes in terminology and standards. For mix design purposes the 1988 edition can be used and gives the same results. The chief change is that for Ordinary Portland Cement and Rapid Hardening Portland Cement, read 42.5 N and 52.5 N respectively in the later edition.)

The description given here is intended to illustrate the principles of the method; the reader would be well advised to carry out a number of design exercises in order to become familiar with the design procedure. The DoE method also includes information necessary to design concretes for indirect tensile strength; containing entrained air and using cement replacements. These are not given here, since they are easily followed by reference to the DoE publication once the basic method of design is understood.

The method is divided into sections, which will be taken in turn. The section and sub-section numbers refer to the numbers shown in the standard mix design form (Fig. 3.12).

Section 1: Strength

1.1   Before describing the method of designing for strength it is necessary to understand the meaning of the term 'characteristic strength' (f_c), which is normally used in strength specifications. Concrete, in common with other materials, is inherently variable, due to within-batch variations of materials, between-batch variations, tolerances and possible

Fig. 3.12 *Concrete mix design form. (Crown copyright)*

Concrete mix design form

Job title ....

Stage	Item		Reference or calculation	Values
1	1.1	Characteristic strength	Specified	{ _____ N/mm² at _____ days Proportion defective _____ %
	1.2	Standard deviation	(Fig 3.13)	_____ N/mm² or no data __ N/mm²
	1.3	Margin	(Table 3.10)	(k = _____) _____ X _____ = _____ N/mm²
	1.4	Target mean strength		_____ + _____ = _____ N/mm²
	1.5	Cement type	Specified	OPC/SRPC/RHPC
	1.6	Aggregate type: coarse Aggregate type: sand		Crushed/uncrushed Crushed/uncrushed
	1.7	Free-water/cement ratio	(Table 3.11/Fig 3.14)	_____ } Use the lower value [ ]
	1.8	*Maximum free-water/cement ratio*	Specified	_____
2	2.1	Slump or Vebe time	Specified	Slump_____ mm or Vebe time _____ s
	2.2	Maximum aggregate size	Specified	_____ mm
	2.3	Free-water content	(Table 3.12)	_____ [ ] kg/m³
3	3.1	Cement content		_____ ÷ _____ = _____ kg/m³
	3.2	*Maximum cement content*	*Specified*	_____ kg/m³
	3.3	*Minimum cement content*	*Specified*	_____ kg/m³
				use 3.1 if ≤ 3.2 use 3.3 if > 3.1 [ ] kg/m³
	3.4	Modified free-water/cement ratio	_____	[ ]
4	4.1	Relative density of aggregate (SSD)		_____ known/assumed
	4.2	Concrete density	(Fig 3.15)	_____ kg/m³
	4.3	Total aggregate content		_____ — _____ — _____ = _____ kg/m³
5	5.1	Grading of sand	Percentage passing 600 µm sieve	_____ %
	5.2	Proportion of sand	(Fig 3.16)	_____ %
	5.3	Sand content	} __ _____ { _____ X _____ = [ ] kg/m³
	5.4	Coarse aggregate content		_____ — _____ = [ ] kg/m³

Quantities	Cement (kg)	Water (kg or *L*)	Sand (kg)	Coarse aggregate (kg)		
				10 mm	20 mm	40 mm
per m³ (to nearest 5 kg)	_____	_____	_____	_____	_____	_____
per trial mix of _____ m³	_____	_____	_____	_____	_____	_____

Items in italics are optional limiting values that may be specified

OPC = ordinary Portland cement; SRPC = sulphate-resisting Portland cement; RHPC = rapid-hardening Portland cement. SSD = based on a saturated surface-dry basis.

errors in batching, mixing, compaction, testing and so on. Hence statistical terminology must be used.

> 'Characteristic strength' implies a strength with a permissible percentage of failures.

For example, a characteristic 28-day compressive strength of 20 N/mm^2 with 5 per cent failures would imply that 95 per cent of results need to be at or higher than the strength level of 20 N/mm^2. The first step in mix design is to find the target mean strength f_m which would be required to give the stated characteristic strength, f_c. Clearly, the target mean strength will need to be higher than the characteristic strength since half of the results will be expected to fall below the former. The increment which must be applied depends on two factors:

(a) *The percentage failures allowable.*
If, for example, only 1 per cent of cube strength failures were permitted at the stated characteristic strength, the target mean strength would need to be higher than if 5 per cent failures were allowed. These percentage failure rates are taken account of by k factors (see the final values in Table 3.10). Smaller percentage failure rates lead to larger k factors since a greater margin between characteristic and target mean strength will be needed.

Table 3.10 k *factors used in statistical control*

Percentage of failures	k factor
16	1.00
10	1.28
5	1.64
2	2.05
1	2.33

1.2 (b) *The variability of the concrete*
If, due to poor quality control, the concrete is more variable, an extra margin will need to be built into the design to compensate. The variability of concrete strength is represented by its standard deviation (s), which will be defined later. To allow for concrete variability, the s value should be known and this, of course, requires that some cube test results be available (usually at least 40 for the s value to be reliable). Where concrete production is starting for the first time or a new type of mix is to be used, standard deviations cannot be known, since no results will be available; hence, in these circumstances, an s value must be assumed and this is taken to be relatively high in order to 'play safe'. Figure 3.13 gives initial values suggested by the DoE method (graph A), together with minimum s values (graph B) in order to guard against the dangers of taking too small a value of s, once results are available. Note that, to some extent, s values reduce as characteristic strength decreases, since the variability of weaker concrete would be expected to reduce (negative strengths are impossible).

1.3 The increment to be added to the characteristic strength f_c is the product of the two terms described: the k factor and the standard deviation s. Hence:

Fig. 3.13 *Relationship between standard deviation and characteristic strength. (Crown copyright)*

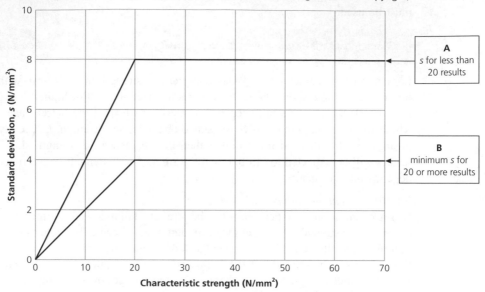

Table 3.11 *Compressive strengths of concrete mixes*

Free-water cement ratio of 0.5

Type of cement	Type of coarse aggregate	Compressive strength (N/mm²) by age (days)			
		3	7	28	91
Ordinary Portland (42.5 N) or sulphate-resisting PC (SRPC)	Uncrushed	22	30	42	49
	Crushed	27	36	49	56
Rapid-hardening PC (52.5 N)	Uncrushed	29	37	48	54
	Crushed	34	43	55	61

1.4
$$f_\mathrm{m} \quad = \quad f_\mathrm{c} \quad + \quad ks$$
target mean strength   characteristic strength

The product *ks* is often referred to as the 'current margin'.

1.5, 1.6 It has already been explained that the strength of the concrete is affected by the free-water/cement ratio, the aggregate type and the cement type (as well as age). Hence the remaining part of this section involves determining the free-water/cement ratio.

1.7 First, the strength appropriate to a free-water/cement ratio of 0.5 and the materials specified is obtained from Table 3.11. This strength is then located on the graph of

Fig. 3.14 *Relationship between compressive strength and free-water/cement ratio. (Crown copyright)*

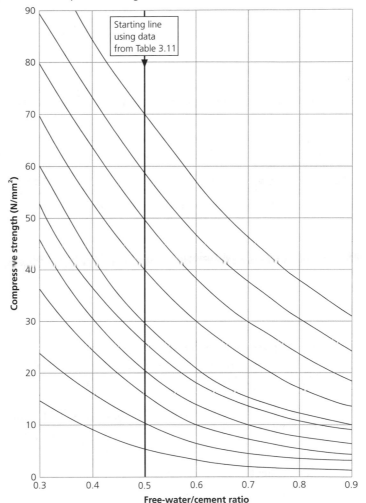

Fig. 3.14, using the 0.5 water/cement ratio starting line. Note that, since curves on this figure are separated by strength increments of 5 or 10 N/mm² at the 0.5 free-water/cement ratio line, they will not generally correspond exactly to the strengths given in Table 3.11. To find the free-water/cement ratio corresponding to the required target mean strength, move parallel to the nearest two lines on Fig. 3.14 (unless the strength happens to fall on a line) from the point located, towards higher or lower strength, as required, until the required target mean strength is obtained (reading horizontally). At this point, the correct free-water/cement ratio is obtained by reading vertically downwards. Since this procedure may require some practice, it is worth checking the result for accuracy. The water/cement ratio obtained should be checked against any maximum value required for durability and the lower of the two values taken.

Section 2: Workability

2.1   Workability is specified by the slump or Vebe time of the fresh concrete.

2.2, 2.3   From the value required and the aggregate shape and maximum size, the water content per cubic metre of concrete is obtained (Table 3.12).

Table 3.12 *Workability and approximate free-water content*					
Slump (mm)		*0–10*	*10–30*	*30–60*	*60–180*
Vebe (seconds)		*>12*	*6–12*	*3–6*	*0–3*
Maximum size of aggregate (mm)	*Type of aggregate*	*Water content (kg/m³)*			
10	Uncrushed	150	180	205	225
	Crushed	180	205	230	250
20	Uncrushed	135	160	180	195
	Crushed	170	190	210	225
40	Uncrushed	115	140	160	175
	Crushed	155	175	190	205

Section 3: Cement content

3.1   It will be evident that, since the free-water/cement ratio and the water content per cubic metre are now known and

$$\text{cement content per cubic metre} = \frac{\text{water content per cubic metre}}{\text{free-water/cement ratio}}$$

the cement content is easily calculated.

3.2   A maximum cement content may be specified (say, to limit heat of hydration) in which case the cement content from subsection 3.1 will need to be checked to ensure that it is not in excess of this value.

3.3   Similarly, a minimum cement content may be specified (say, for durability), so that the figure from subsection 3.1 must be in excess of any such value. If it is not, the value specified must be used and the modified free-water/cement ratio calculated.

Section 4: Fresh concrete density

This is important as a part of the design process because it affects the yield of the concrete. The DoE method uses a graphical system for calculation of fresh density; an alternative means is given after this explanation of the DoE method.

4.1, 4.2   If the relative density of the aggregate (usually based on the saturated surface dry, SSD, condition) is known, the fresh density can be related to the water content of the mix, as in Fig. 3.15.

Fig. 3.15 *Estimated wet density of fully compacted concrete. (Crown copyright)*

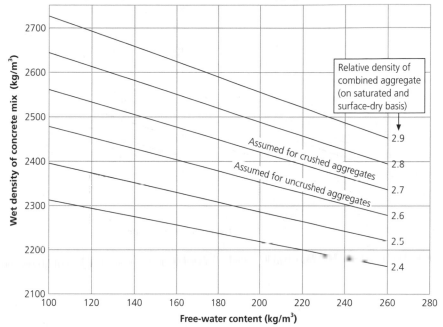

4.3 The aggregate content per cubic metre of concrete is then:

fresh density – (cement content + water content)

Section 5: Proportion of sand

5.1, 5.2 The proportion of sand to total aggregate depends on the grading of the sand as indicated by the proportion of sand passing the 600 μm sieve. It also depends on the maximum aggregate size, the workability and the free-water/cement ratio. (A smaller maximum aggregate size, higher workability and a higher water/cement ratio each increases the sand requirement.) The DoE method gives a proportion of sand corresponding to 'normal' cohesion. If higher or lower values are required, the percentage given could be increased or decreased by, say, 5 per cent respectively. Note that there are 12 sets of graphs in Fig. 3.16 – it is important to use the one appropriate to the chosen aggregate size and workability.

5.3 Sand content = percentage of sand × total aggregate content.

5.4 Finally, the coarse aggregate content is found by subtracting the sand from the total aggregate.

Worked example

Produce batch quantities for 1 cubic metre of concrete to have a slump of 30 to 60 mm and a characteristic 28-day compressive strength of 25 N/mm² (5% failures permitted). No previous results are available. The following materials are to be used:

Fig. 3.16 *Recommended proportions of sand according to percentage passing a 600 µm sieve. (Crown copyright)*

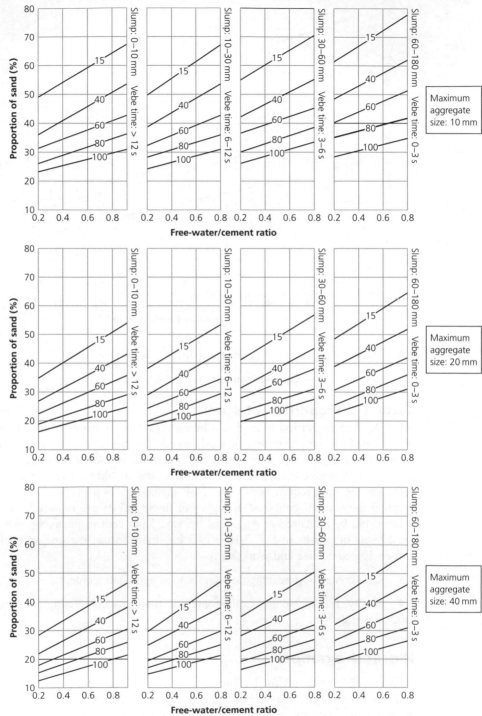

Cement: Ordinary Portland (42.5 N)

Sand: Crushed, relative density 2.7; 55 per cent passes the 600 μm sieve

Coarse aggregate: Crushed, max. size 20 mm, relative density 2.7

This example has been worked on the standard design form in Fig. 3.17. Note:

1. The figure obtained from Table 3.11 is 49 N/mm^2, corresponding to 42.5 N cement, crushed aggregate and 28 days strength specification. This figure is used as the starting point on Fig. 3.14.
2. Provision is made for aggregates to be divided into single sizes in the final summary of Fig. 3.17. This is only used for high-quality concrete and has not been carried out in the worked example. Where this is required the aggregates can be divided as follows:

    1 : 2   for 20 mm maximum size in 10 and 20 mm stockpiles

    1 : 1.5 : 3   for 40 mm maximum size in 10, 20 and 40 mm stockpiles

Trial mixes

It is most important to appreciate that no design method can produce exactly the right combination of fresh and hardened properties every time; a trial mix should always be made to ensure that the properties are as required with the particular materials and equipment used. The figures used in the design are based on 1 cubic metre of concrete, which would be excessively large for a trial mix. Hence the form of Fig. 3.12 provides for the calculation of what would normally be smaller batch quantities, for example, 50 litres for a trial mix. To obtain batch quantities for 50 litres (0.05 m^3), the batch quantities for 1 cubic metre would be multiplied by 0.05.

Concrete which is produced by the trial mix would be tested for workability and fresh density. Cubes would then be made to check for strength. If workability or strength are slightly in error, corrections can be made as follows:

*   **Workability**

    The correct water content for workability can often be judged by experience. If, for instance, in the worked example given, the actual water per cubic metre is found to be 220 kg/m^3, the cement content must be increased to keep the water/cement ratio and hence the strength constant. Hence the new cement content would be:

    $$\frac{220}{0.58} = 379 \text{ kg}$$

    The fresh density (if not measured) would also be expected to reduce slightly. Figure 3.15 indicates that the correct value would be approximately 2390 kg/m^3. The new total aggregate content is therefore

    $$2390 - (220 + 379) = 1791 \text{ kg}$$

    The sand content would be found as previously.

Fig. 3.17 *Concrete mix design form.*

Concrete mix design form

Job title .... *Worked Example*

Stage	Item		Reference or calculation	Values

1

1.1 Characteristic strength — Specified — _25_ N/mm² at _28_ days
Proportion defective _5_ %

1.2 Standard deviation — (Fig 3.13) — _____ N/mm² or no data _8_ N/mm²

1.3 Margin — (Table 3.10) — (k = _1.64_) _1.64_ × _8_ = _13_ N/mm²

1.4 Target mean strength — _25_ + _13_ = _38_ N/mm²

1.5 Cement type — Specified — OPC/SRPC/RHPC

1.6 Aggregate type: coarse — Crushed/uncrushed
Aggregate type: sand — Crushed/uncrushed

1.7 Free-water/cement ratio — (Table 3.11/Fig 3.14) — _0.58_
1.8 *Maximum free-water/cement ratio* — *Specified* — Use the lower value | 0.58 |

2

2.1 Slump or Vebe time — Specified — Slump _30–60_ mm or Vebe time _____ s

2.2 Maximum aggregate size — Specified — _20_ mm

2.3 Free-water content — (Table 3.12) — | 210 kg/m³ |

3

3.1 Cement content — _210_ ÷ _0.58_ = _362_ kg/m³

3.2 *Maximum cement content* — *Specified* — _____ kg/m³

3.3 *Minimum cement content* — *Specified* — _____ kg/m³

use 3.1 if ≤ 3.2
use 3.3 if > 3.1 — | _____ kg/m³ |

3.4 Modified free-water/cement ratio _____ | _____ |

4

4.1 Relative density of aggregate (SSD) — _2.7_ known/assumed

4.2 Concrete density — (Fig 3.15) — _2400_ kg/m³

4.3 Total aggregate content — _2400_ – _210_ – _362_ = _1828_ kg/m³

5

5.1 Grading of sand — Percentage passing 600 μm sieve _55_ %

5.2 Proportion of sand — (Fig 3.16) _36_ %

5.3 Sand content — _0.36_ × _2400_ = | 864 kg/m³ |
5.4 Coarse aggregate content — _1828_ – _864_ = | 964 kg/m³ |

Quantities	Cement (kg)	Water (kg or *L*)	Sand (kg)	Coarse aggregate (kg)		
				10 mm	20 mm	40 mm
per m³ (to nearest 5 kg)	360	210	865	___	965	___
per trial mix of ___ m³	___	___	___	___	___	___

Items in italics are optional limiting values that may be specified

OPC = ordinary Portland cement; SRPC = sulphate-resisting Portland cement; RHPC = rapid-hardening Portland cement. SSD = based on a saturated surface-dry basis.

- **Strength**

  The point on Fig. 3.14 corresponding to the strength obtained and the water/cement ratio used should be located. Then proceed parallel to the curved lines until the desired strength is obtained and read the new water/cement ratio. If, for instance, in the worked example, a strength of 35 N/mm^2 was obtained instead of 38 N/mm^2, find the point corresponding to 35 N/mm^2 and 0.58 water/cement ratio. Move upwards parallel to the curved lines until 38 N/mm^2 is reached and then read vertically. This produces a water/cement ratio of 0.55. The remaining mix details are now recalculated.

Where both workability and strength are incorrect, a new water content and water/cement ratio would be obtained and the method completed as before.

Second trial mixes should only be necessary where substantial errors are involved, though a final full-scale trial mix is advisable before going into production. Adjustments for moisture in aggregates are, of course, essential if trial mixes are to fulfil their function.

Calculation of concrete fresh density

It is only from an accurate knowledge of concrete density that batch masses of cement and aggregates for a given volume of concrete can be reliably predicted. The following technique is therefore given and serves as an alternative to the DoE method of mix design.

Suppose, for example, that 400 m^3 of concrete are to be produced to the specification given in Fig. 3.17. The batch quantities per cubic metre are:

Water	210 litres
Cement	360 kg
Sand	865 kg
Coarse aggregate	964 kg.

The relative densities of the aggregates are required. The relative density of the cement is normally assumed to be 3.15.

The figures are tabulated as in Table 3.13 and, to simplify them, all figures are divided by the mass of cement, giving masses which would correspond to 1 kg of cement.

Table 3.13 Example calculation of fresh density

Material	Mass (kg)	Mass ratio (per kg of cement)	Rel density (SSD)	Volume (litres)
Water	210	0.583	1.00	0.583
Cement	360	1	3.15	0.317
Sand	865	2.403	2.70	0.890
Coarse aggregate	965	2.681	2.70	0.993
Total	2400	6.667		2.783

The volumes of each component in litres are easily obtained by dividing the reduced mass of the component by its relative density. Then:

$$\text{concrete density} = \frac{\text{total mass}}{\text{total volume}}$$

$$= \frac{6.667}{2.783}\,\text{kg/litre}$$

$$= 2.396\ \text{kg/litre}$$

or, to nearest 10 kg/m^3 = 2400 kg/m^3

(This figure agrees with the value obtained from Fig. 3.15.) To obtain the batch quantities for 400 m^3 of concrete, the batch quantities per m^3 are simply multiplied by 400. If, as may be the case, mass ratios only are given, the ratios are multiplied by the appropriate factor:

$$\frac{2400}{6.667} \quad \text{(in our example)}$$

to give batch quantities per cubic metre. The procedure is then as before.

If the presence of air is to be allowed for in calculating density (as, for example, in air-entrained concrete, used to increase frost resistance), this is easily done by increasing the total volume of concrete (but not the mass) by the percentage of air that is expected.

3.11 Curing of concrete

Curing is the process by which cement is enabled to hydrate. Since water is involved in the process it is important to maintain the concrete in a saturated condition for a sufficient length of time to develop a satisfactory hydrate structure. This is essential not only in relation to strength development but also to enable the concrete to withstand the stresses caused by shrinkage (see page 83). The minimum period for this is usually considered to be about one week, though it may be longer with cements which hydrate slowly, such as pozzolanic types. The temperature is also an important factor, the 'ideal' value being between 5 and 15 °C since hydration is very slow at lower temperatures, while high temperatures tend to cause evaporation, together with an inferior 'gel' structure. With care, concrete can be cured up to temperatures of about 30 °C. In no circumstances must concrete be allowed to freeze before it reaches maturity; the accompanying expansion of water could cause serious surface damage.

As an aid to memory it may be added that if good concrete is to be produced, the four 'C' s should be considered:

Cement content
Cover to steel reinforcement
Compaction
Curing

3.12 Statistical interpretation of cube results

Following the section on the design of concrete, we return to the subject of variability. The reasons for strength variations have already been given and it will be evident that the extent of variations will depend to a large degree on the quality control procedures carried out on any one site. Figure 3.18 shows, for example, two histograms relating to 50 cube tests on two sites A and B. It will be evident that the curve A, although of lower mean strength, is much narrower than that of curve B and that, with site A, unlike site B, there are no cube results below 30 N/mm². The curves have been drawn such that characteristic strengths based on 8 per cent failures are the same, although site A satisfies this requirement with a lower mean strength (hence a more economical concrete) and fewer very low cube results.

The standard deviation (*s*) of a set of *n* results, values $x_1, x_2, \ldots, x_i, \ldots, x_n$ of average value $\bar{x}$ is given by

$$s = \sqrt{\frac{\sum (x_i - \bar{x})^2}{n-1}}$$

Fig. 3.18 *Histograms drawn from the results of two sets of 50 concrete cube tests. Graph A represents good-quality control and graph B poor-quality control. The total area under each graph is the same. Also, the area under each curve to the left of the 35 N/mm² line is the same, implying equal numbers of results beneath this strength. Curve B is, however, indicative of a greater probability of very low cube results than curve A.*

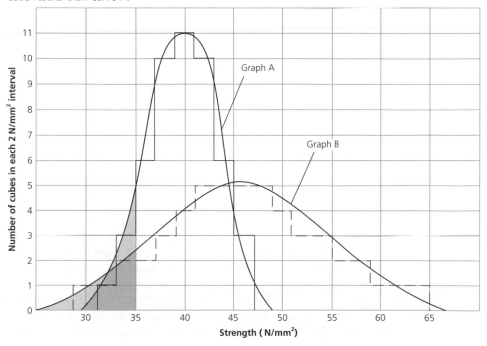

or, to put it alternatively,

$$s = \frac{\sqrt{\text{sum of squares of (each individual result minus mean strength)}}}{n-1}$$

Variability can never be eliminated. Successive cubes from the same batches of cement and aggregates will produce a standard deviation of about 2 N/mm^2 due to within batch variations and equipment tolerances. The value of s obtained in practice for ordinary concrete of mean strength 40 N/mm^2 might vary from 4 N/mm^2, which represents good quality control, to 8 N/mm^2, which represents poor quality control.

Perhaps the most direct way of checking that the characteristic strength is satisfactory is to keep calculating the mean strength (f_m) and standard deviation (s) using, say, the most recent 50 results and then to calculate characteristic strength (f_c) by the equation

$$f_c = f_m - ks$$

k being the factor appropriate to the permissible number of failures, as given in Table 3.10. Such a method would, however, be mathematically quite arduous and would indicate rather slowly any change in quality control. It is especially important that any adverse change in strength be readily detectable as soon as possible and one convenient technique is the use of simple quality control charts in which strengths of consecutive cubes are represented as points. It is often found more convenient in practice to take means of four cubes and to plot the means, since this reduces the variability of results about the mean strength.

Suppose, for example, that the required characteristic strength is 30 N/mm^2 with 5 per cent failures allowed and standard deviation equal to 6 N/mm^2. The target mean strength would then be

$$f_m = f_c + ks$$

$$= 30 + 1.64 \times 6$$

$$= 39.84 \text{ (say 40) N/mm}^2$$

If means of 4 are plotted, the standard deviation of the means should be

$$\frac{\text{standard deviation of individual results}}{\sqrt{\text{number in group}}} = \frac{6}{\sqrt{4}} = 3 \text{N/mm}^2$$

Hence, not more than 5 per cent of means of 4 should be below

$$40 - 1.64 \times 3 = 35.08 \text{ (say 35) N/mm}^2$$

A chart could then be constructed, as in Fig. 3.19, with mean strength 40 N/mm^2 and 'action lines' at 35 and 45 N/mm^2. If there is a definite trend towards, say, the 35 N/mm^2 line, then action, in the form of increasing the mean strength of the concrete, must be taken, since the points should be evenly spaced about the means with no more than one result in twenty (5 per cent) below the lower line if concreting is proceeding satisfactorily. 'Warning' lines may also be included corresponding to, say, 10 per cent limits. Since, for 10 per cent failures, k is 1.28 (Table 3.10), the lines would be at strengths of

Fig. 3.19 *Simple control chart for monitoring progress during production of concrete. In this case progress is satisfactory since there are no results outside 'Action' limits and results are equally distributed about the mean.*

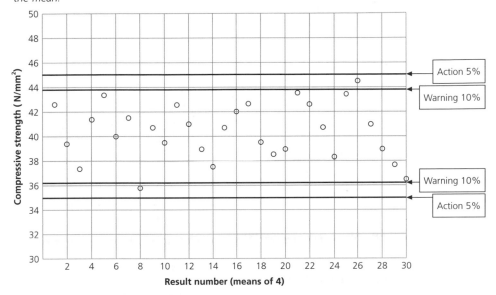

$40 \pm 1.28 \times 3 = 43.8$ and 36.2 N/mm² approximately.

The warning lines would serve to indicate that the concrete quality is varying. Increases in mean strength should also be checked, since unnecessarily high mean strengths (or reductions in standard deviation) may allow a reduction in the cement content of the concrete, hence saving cost. Small corrections to strength are, in fact, most commonly made by small alterations to the cement content.

A simple guide is that, **to increase mean strength by 0.8 N/mm², the cement content should be increased by 5 kg/m³ of concrete** and vice versa. The aim, in all concrete testing is to obtain results as soon as possible – accelerated curing may be used to predict 28-day strength as early as one day after placing provided reliable correlations have been produced. Hence, faulty concrete can be identified and removed before it is fully mature or further lifts have been cast. More elaborate charts may also be used; for instance, a second chart plotting variability within groups of 4 cube results, compared with a 'target mean variability'. This would reveal any change in quality control (represented by standard deviation) as work proceeds.

3.13 Shrinkage and moisture movement of concrete

Drying shrinkage is the contraction occurring when concrete is dried for the first time.

This should be distinguished from subsequent movements resulting from moisture changes and referred to as 'moisture movement' because the initial shrinkage is partly irreversible

Fig. 3.20 *Drying shrinkage and moisture movements in concrete.*

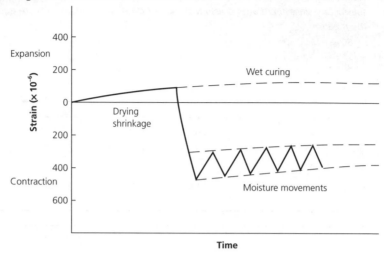

and is much more likely to result in damage to the concrete than any subsequent movements. Figure 3.20 gives a schematic representation of typical volume changes of concrete with time. During wet curing there is a slight expansion – indeed, it should be noted that cement expands on hydration by over 100 per cent, filling at least part of the space previously occupied by water. Shrinkage occurs when, on drying, water from within the cement 'gel' is removed, causing gel surfaces to approach slightly. Hence, the origin of shrinkage lies within the hydrated cement and it follows that rich, wet mixes, which contain more cement gel, shrink more. Long-term shrinkage strains vary from 200×10^{-6} for lean, low-workability mixes, to over 700×10^{-6} for rich, wet concretes.

(Note that 1×10^{-6} is equivalent to 1 mm per kilometre movement – a very small movement but important if it throws concrete into tension.)

Approximately 40 per cent of this movement is irreversible – the concrete does not regain this proportion if rewetted. Moisture movement does, however, occur on each subsequent wetting and drying, as water is admitted to and removed from the cement gel. Even the lower shrinkage value given above could cause tensile failure in the concrete if shrinkage is restrained or if differential shrinkage occurs between different parts of a structure since the tensile strain capacity of mature concrete is only about 100×10^{-6} and that of young concrete is much lower still. Hence, it is important to cure concrete until it has some ability to withstand the strains occurring due to shrinkage. Examples of situations in which shrinkage must in any case be allowed for are:

- Ground floor slabs, hard standings and roads, where subgrade friction can generate stresses sufficient to cause failure unless joints are provided.
- Mass concrete, where surface shrinkage is restrained by inner layers. Surface shrinkage reinforcement can be provided to control this.
- Clay brickwork infill panels in reinforced concrete frames – especially if the bricks are fresh from the kiln, leading to a risk of expansion. Movement provision is essential.

The concrete frames of reinforced concrete buildings do not normally suffer greatly from shrinkage damage since their shrinkage is largely unrestrained.

3.14 Durability of concrete

In the 1980s, there was increasing concern over the rate of deterioration of some concrete structures. This, and the fact that concrete buildings can be very expensive to repair, together with a revitalised steel industry, led to a decline in the use of concrete for some applications. In most instances, deterioration of concrete could be linked to failure to implement existing technology due, for example, to a lack of supervision, though, in a few instances, decay has been the result of inadequate understanding of the behaviour of the material. There is now a greater awareness of potential problems and greater confidence in the performance of concrete generally.

Almost all forms of deterioration are the result of water ingress.

The need to keep water out of structures, or to take special care where this is not feasible, cannot be overstated. The action of water may be two-fold:

1. **Physical attack,** where water itself crystallises on freezing or where drying out causes damage due to crystallisation of dissolved salts in the water.
2. **Chemical damage.** Most forms of chemical damage are associated with water since water is essential for the formation of ions – the chemically active form of salts or other materials.

The main types of deterioration are given below.

Frost attack

There are three prerequisites for frost attack:

1. A permeable form of concrete able to admit water.
2. The presence of water (though materials need not be saturated to be affected by frost).
3. Temperatures below 0 °C.

The mechanism of attack has long been attributed to the 10 per cent expansion of absorbed water on freezing, though it is now known that further damage can result from *movement* of water within concrete on cooling below 0 °C. Ice builds up in large pores and cracks and may cause very large expansion in such positions, while other areas become dryer (desiccation). Perhaps the extreme example of this is frost heave, as found in soils where large ice lenses can be obtained. Frost damage in concrete is exacerbated as follows:

1. Horizontal surfaces tend to absorb more water in wet conditions, take a longer time to dry and also cool quicker by radiation to the sky on a clear night.
2. Very low temperatures increase the extent of migration of water and result in freezing to greater depths in the concrete.

3. Repeated freezing and thawing (rather than prolonged steady freezing) add to the damage. Once freezing has occurred no further damage is done at a steady low temperature. However, a thaw followed by a further frost will start another cycle of damage. Hence climates such as those found in the northern UK could be more damaging than colder climates where thawing is less frequent and, in any case, precipitation in such climates is more likely to be in the form of snow which, being solid, cannot penetrate and therefore cannot damage concrete until thawing occurs.
4. Weak, permeable concretes which absorb water more readily are much more vulnerable to attack.
5. De-icing salts melt the ice but add to saturation of the concrete. They then crystallise on drying, adding to the damage.

Most of the measures required to avoid frost attack will be evident from the above, though it should be added that use of air-entraining agents has been found to be particularly effective in avoiding damage to surfaces at risk. It is important to appreciate that new concrete with its incomplete hydrate structure and high permeability is particularly at risk from frost, and must be protected by an insulating material until it has sufficient maturity to resist frost.

Sulphate attack

Sulphates are often found in the ground, mainly associated with clay soils. These have low permeabilities, which means there is little chance of sulphates being washed out. To be active the sulphates must be in solution so that the risk of attack depends not only on the salt content of a soil but on the presence and movement of moisture. The principal mechanism is one in which sulphate ions react with hydrated C_3A in the cement, producing an expansive product, calcium sulphoaluminate, which disrupts the concrete. Sulphates commonly occur in three forms – calcium, sodium and magnesium – which pose quite different risks. Calcium has low solubility (about 1.4 mg of ions per litre) and does not constitute a high risk. Of the other two, magnesium is more harmful because:

- The reaction product is insoluble, precipitating out of solution. This avoids 'chemical congestion' in the solution, which would tend to reduce the rate of attack.
- Magnesium sulphate also reacts with the C_3S hydrate in cement.

When estimating the risk of sulphates **it is therefore important to identify the type of sulphates present**. Total sulphate content could be misleading if it were mainly in the form of calcium sulphate. Careful sampling is also necessary since levels may vary widely from place to place and ground water may be diluted during sampling or after wet weather. The best procedure is normally analysis of ground water or water obtained from a 2 : 1 water–soil mixture.

For ground water containing up to 1.4 g/litre of SO_4 ions, OPC can be used provided a cement content of at least 330 kg per cubic metre is used with a water cement ratio not higher than 0.5.

At sulphate levels up to 6.0 mg/litre without high magnesium concentrations, use of pozzolanas at high replacement levels, or sulphate-resisting Portland cement, would be necessary.

At higher sulphate levels, or when high magnesium levels are found, the use of sulphate-resisting Portland cement in conjunction with a waterproof coating would be essential.

In all cases it is of paramount importance that attention is given to achieving full compaction of a good-quality impermeable concrete.

Crystallisation damage

This may occur in concretes which are resistant to chemical attack. Sulphates (or other salts) may be admitted by permeable concretes and if, at a later stage, drying occurs, these salts crystallise, causing damage similar to that caused by frost. The condition may be severe when concrete is subject to wetting and drying cycles – as, for example, in sea walls around the high-tide mark. Attack may also be serious if one area of a structure is saturated while an adjacent area is dry – salts migrate towards the dry area, causing extensive crystallisation in the region of drying. This type of damage may also be avoided by the use of low water/cement ratio concretes, fully compacted to produce low permeability.

Alkali–silica reaction (ASR)

Although relatively uncommon at the moment ASR, sometimes called 'concrete cancer', has evoked a strong public response because it is a relatively new form of deterioration and there is no effective repair technique for structures that are seriously affected. ASR occurs when three conditions are met:

- The concrete has a very high pH value (over 12) resulting from the presence, in quite small quantities, of sodium and potassium alkalis originating from the oxides of these elements if present in the cement (high-alkali cements).
- Aggregates contain reactive forms of silica such as found in opal, tridymite, some flints and cherts.
- The presence of water.

The reactive components of the aggregates form a gel, leading to expansion and disruption of the concrete, often producing 'map' cracking (Fig. 3.21). The reaction is slow and the cracking usually takes 10–20 years to become evident. There is a 'pessimum' level of reactive aggregate of about 2 per cent since smaller quantities do not form sufficient gel while larger quantities 'swamp' the excess alkali present. Once formed, cracks admit further water, accelerating the rate of deterioration. ASR can be avoided by applying one or more of the following:

- *Use of low-alkali cements.* For example, low alkali versions of OPC (BS 12) containing less than 0.6 per cent alkali, expressed as Na_2O equivalent. Alternatively, pozzolanic additions are known to be beneficial in reducing ASR.

Fig. 3.21 *Map cracking due to ASR in concrete. This is sometimes called 'Isle of Man' cracking because cracks meet in groups of three.*

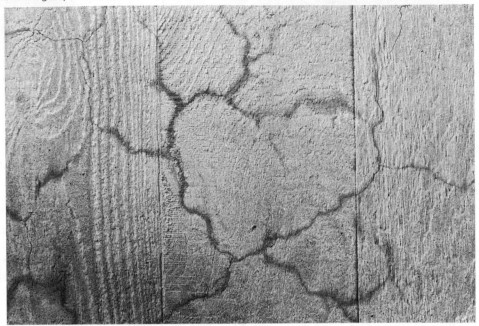

- *Avoidance of reactive aggregates.* Some types, such as granite or limestone, are unlikely to be reactive. In areas of doubt, expansion tests can be carried out on samples of concrete made from the aggregates in question placed in water at 38 °C for some weeks.
- *Exclusion of water.* Most structures affected to date have been severely exposed; for example, concrete bridges or buildings near the sea. The risk to interior concrete is very small except in swimming pools or other damp areas. Once attack has commenced, it may be very difficult to exclude further water from affected structures.

The types of attack referred to above are summarised in Table 3.14.

3.15 Corrosion of steel reinforcement

This is one of the commonest problems in structural concrete. Where damage is localised, repair is possible, though in view of the labour intensive nature of the repair process, it can be very expensive. Extensive corrosion of reinforcement may result in the need to demolish the structure.

The corrosion of steel in moist environments is described elsewhere (p. 205). Steel in concrete will be protected, provided an alkaline environment exists, since a protective layer of Fe_2O_3 (ferric oxide) is formed and there is no tendency to ionise (corrode). Such an environment is provided by calcium hydroxide, one of the products of hydration. This material is, however, attacked from the surface by carbon dioxide in normal moisture-containing atmospheres, producing almost neutral calcium carbonate:

Table 3.14 *Common modes of deterioration of concrete*

Type of attack	High risk situations	Symptoms	Means of avoidance
Frost	Poor quality concrete; exposed horizontal surfaces	Surface scaling/ disintegration	Use concrete of suitable specification for exposure category; protect new concrete; avoid ponding of water; use air-entraining agents
Sulphates	Clay soils, below water table especially if soluble sulphates present	Cracking/ disintegration of concrete	Use resistant cement at low water/cement ratio; in extreme cases use impermeable coating
Alkali-silica reaction	Exposed structures; high-alkali cements; reactive aggregates	'Map' cracking	Use low-alkali cement; avoid use of reactive aggregates

$$Ca(OH)_2 + CO_2 \longrightarrow CaCO_3 + H_2O$$

The rate of attack and hence the depth of penetration of this 'carbonation' reaction depend chiefly on the permeability of the concrete. Well-compacted, low water/cement ratio concretes have low permeability such that carbonation would only reach perhaps 5 mm depth – much less than the depth in normal steel reinforcement over many years. In lower quality concretes, much higher carbonation depths of 25 mm or more may occur, putting steel at risk. The need to fix steel accurately in position at the time of construction in order to give *consistently* adequate cover will be apparent.

The steel must be fixed at a depth greater than the 'minimum' depth by a margin which is sufficient ensure adequate cover at all positions.

The risk of corrosion to steel in existing structures can easily be checked by breaking off pieces of concrete from the surface and spraying the freshly exposed surface with a phenolphthalein solution. This indicator turns pink in alkaline areas, revealing clearly the carbonation depth (Fig. 3.22). (The damaged area must, of course, be made good after the test.) Depths of reinforcing bars can be ascertained by using a covermeter, based on the magnetic effect of steel. Unless the steel is well beneath the carbonated layer, there is a risk that corrosion could occur.

Where chlorides are present – for example, from admixtures used in the concrete, or de-icing salts – the corrosion risk is greatly increased since chloride ions have a de-passivating effect; that is, they reduce the ability of an alkaline environment to protect the steel from corrosion. They also increase the electrical conductivity of the concrete and their use is now strictly controlled in structural concrete.

Fig. 3.22 *Phenol phthalein test on precast concrete post used in the prefabricated housing; the alkaline areas are darker.*

In situations of high risk the likelihood of reinforcement corrosion can be reduced by:

- Utilising anti-carbonation coatings (usually also decorative) on the concrete surface to reduce the rate of carbon dioxide ingress.
- Using special steels for reinforcement; for example, in order of increasing cost, galvanised, epoxy coated and stainless steel.

The best means of achieving longevity, nevertheless, remains the use of good-quality concrete with careful supervision to ensure that cover to steel is always adequate.

3.16 Admixtures (BS 5075)

Admixtures are materials other than cement, water or aggregates added at the mixer.

The use of admixtures has increased greatly in recent years and there are many instances in which marked improvements in properties for specific applications are obtainable. However, careful precautions regarding their use are advised for the following reasons:

- Doses are usually very small, often less than 1 per cent by weight of cement, hence, careful batching is necessary, the use of special dispensers being recommended.
- Effects are sometimes very sensitive to variations in dose; hence, good supervision is essential.
- Side effects are often present, especially if overdoses occur.

- To ensure effective dispersion, liquid admixtures should be added to the mixing water rather than to the mixer and water-soluble solid admixtures should be dissolved in the mixing water before use.

There are four main categories of admixture, though in practice it may be convenient to combine several effects within one admixture. In particular, most admixtures are formulated to have a plasticising action since very little increase in cost is needed to achieve this. Most admixtures are now manufactured to BS 5075, and with careful use can be very beneficial to concrete for specific applications.

Water reducers (plasticisers)

These are perhaps the most widely used admixtures in the concrete industry, their use in ready-mixed concrete being especially common. The main water-reducing compounds are lignosulphonates and hydrocarboxylic acids and they operate by attaching themselves to cement grains, imparting a negative charge which causes grains to disperse more effectively. Hence, the formation of 'flocs' (groups of cement particles), which tend to trap mixing water, is avoided (Fig 3.23) Water reducing admixtures may be used for any of the following reasons:

Fig. 3.23 *Cement particles, on account of surface charges, form into 'flocs', trapping water and reducing workability.*

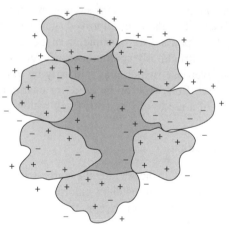

1. To allow a water reduction, hence reducing the water/cement ratio and increasing the strength.
2. To produce an increase in workability at a given water and cement content.
3. To allow a reduction in water and cement contents in concrete of a given strength, hence reducing the cost.

Typical water reductions are in the region of 5–15 per cent when used in items 1 or 3 above. Side effects include air entrainment and retardation of set, though these are only slight at normal doses. Plasticisers are relatively cheap and form an economical means of improving concrete quality if used in 1 above. However, when used in 3 above, the reduced cement content could lead to durability problems.

More recently, a number of 'superplasticisers', based on compounds such as sulphonated melamine formaldehyde, have been introduced. These can be used at higher dosages than ordinary water reducers without the same side effects. Since they are quite expensive, they are used mainly for applications 1 and 2 above. In the latter, the concrete, which has a slump of over 150 mm, is often referred to as 'flowing concrete', since it is more or less self-levelling and self-compacting, reducing labour costs in applications such as large floor slabs. The effect of superplasticisers is short lived – only 30–45 minutes, hence they are usually added to the concrete just before placing. Care is necessary to avoid segregation but hardened properties, such as strength, shrinkage and creep, do not appear to be adversely affected.

Retarders

The object of these would normally be to retard the set (rather than the strength development) of the concrete. Hence, their effect should last only a few hours. They are normally based on lignosulphonates, hydrocarboxylic acids or sugars. They act mainly on those compounds in cement which are responsible for setting and early strength; that is, C_3A and C_3S. Uses include the following:

- To reduce the setting rate when concreting in hot weather. In many cases the retarder can improve long-term strength when used in this way, since slower hydration of cement leads to an improved 'gel' structure.
- In very large pours – for example, in mass concrete foundations in order to produce monolithic (structurally continuous) concrete despite a prolonged placing period.
- A large overdose (which may still only be 0.5 per cent by weight of cement) will 'kill' the set. Retarders can be used, for example, to avoid hardening of concrete in the drum of a ready-mix agitator truck which has broken down.

Accelerators

These are designed to accelerate the early strength development of concrete. They are traditionally based on calcium chloride, which increases the hydration rate of the calcium silicates and, to some extent, the C_3A in the cement. Substantial improvements can be obtained between the ages of 1 day and 7 days, especially in cold weather; for example, the 3-day strength of concrete may be doubled at a temperature of 2 °C. Long-term strength is unaffected by the use of calcium chloride. Most important are the side effects of the material.

The susceptibility of steel in structural concrete to corrosion is greatly increased in the presence of chlorides.

BS 5328 limits chlorides, measured as chloride ion content by weight of cement, as follows:

Prestressed concrete, steam-cured concrete	0.1 per cent
Sulphate-resisting cement with embedded metal	0.2 per cent
Other concretes containing steel	0.4 per cent

(These chloride ion contents are equivalent to approximately 50 per cent greater percentages of anhydrous calcium chloride by weight.)

Any chorides present in the aggregates – for example, from sea-dredged sources – must be included in the above figures. The shrinkage of concrete is also increased by the presence of chlorides. Chloride-free accelerators are now available but these are more expensive than choride-based accelerators and it should be noted that more rapid strength development can be obtained by other means; for example, use of water-reducing agents or increased cement content.

Air-entraining agents

These are based on substances such as vinsol resins, which reduce the surface tension of the water, allowing stable air bubbles to be introduced by mixing. These air bubbles are much smaller than pockets of entrapped air; most are less than 1 mm in diameter and do not admit water, even in saturated conditions. The uses of air-entraining agents are based on the following properties of air-entrained concrete:

- Resistance of hardened concrete to frost and de-icing salts is greatly improved. The close distribution of small air bubbles throughout the mortar fraction of the concrete appears to act as a means of pressure relief as freezing occurs in the concrete. Current specifications for concrete roads in the UK stipulate an air content of 5 per cent in the top 50 mm of the concrete (20 mm aggregate).
- Resistance to segregation and bleeding are improved, together with workability, hence air-entrained concrete could be used to improve harsh mixes or to permit placing in conditions where segregation might be a problem. The air present reduces the concrete strength, though at least part of this shortfall can be avoided by the use of lower water contents for a given workability.

3.17 Lightweight concretes

These may be defined as concretes that have a dry density of less than 2000 kg/m^3, and they offer the following advantages over ordinary dense concrete:

- Since less raw material is involved, it may be argued that lightweight concretes help to conserve materials resources.
- A number of lightweight aggregates are produced from what are otherwise waste materials – for example, pulverised fuel ash and blast-furnace slag.
- Foundation loads are reduced, hence foundation size can be decreased.
- Higher lifts can be cast for a given formwork system, since formwork pressures are lower; hence, the number of construction joints can be reduced.
- Some lightweight concretes, such as precast aerated blocks and no-fines concrete, provide an excellent key for rendering and plastering.

- Fixings are often more easily made than with dense concretes. Some lightweight concretes can be nailed (using cut nails).
- Thermal insulation is improved compared with dense aggregate concrete.
- Fire resistance of lightweight concretes is superior to that of most dense concretes.
- High-frequency sound absorption is better than that of dense concrete.
- Lightweight precast blocks are much easier to handle and cut than dense concrete equivalents.

There are three main methods of producing lightweight concretes, each method producing concrete with characteristic properties and applications.

Lightweight aggregate concrete

A large range of concrete densities – from 500 kg/m^3 to 2000 kg/m^3 – can be obtained using lightweight aggregates. The thermal conductivity of concrete decreases as density decreases, values ranging from about 1.0 W/m°C for the higher densities to under 0.2 W/m°C for concrete of very low density. Strengths decrease similarly with density, though certain aggregates may give higher strengths than others at a given density. An aggregate would normally be chosen according to the density and strength required, subject also, of course, to the local availability. Lower density concretes would be used in situations where, perhaps, thermal insulation properties rather than strength are important – for example, as insulating layers in floors or roofs. Concretes having densities of over 1000 kg/m^3 can, however, be used 'structurally' and there are now many structures incorporating reinforced and prestressed lightweight aggregate concrete. By choice of suitable aggregates, such as foamed slag or sintered pulverised fuel ash, concrete strengths of over 40 N/mm^2 are obtainable with densities in the range of 1500–2000 kg/m^3. The substitution of natural sand for lightweight fine aggregate can be used to increase strength still further, although density will also clearly rise. In designing lightweight concrete structures, allowance must be made for the lower elastic modulus and higher shrinkage and creep of lightweight aggregate concretes.

The production of lightweight aggregate concretes requires care because:

1. Design is best tackled from volume considerations in conjunction with trial mixes. Hence, bulk densities must be known if, as is normally the case, weight batching is used.
2. Aggregates often absorb a great deal of moisture: 15 per cent or more by weight. Hence, the effective free-water/cement ratio, an important factor in controlling strength, is often difficult to predict.

Aerated concrete

Aerated concrete comprises a cement/sand mortar into which gas is introduced. This may be obtained chemically, for example, by the use of aluminium powder, which produces hydrogen in the presence of alkalis released by the cement, or by the use of foaming agents. Careful control of plastic properties is essential in order that the gas aerates the mortar without escaping; hence, these concretes are mainly produced in the form of manufactured precast blocks. Densities in the range 400–800 kg/m^3 are obtained, corresponding to compressive strengths of 2–5 N/mm^2.

Conventionally cured aerated concretes would have high shrinkage but this can be reduced by high-pressure steam curing (autoclaving). Hence many commercial precast blocks are autoclaved and these would also normally contain a pozzolanic material, such as pulverised fuel ash, which contributes to strength.

With increased standards of thermal insulation now being mandatory in residential buildings, the use of aerated concretes, which have very low thermal conductivity (in the region of 0.2 W/m °C) has increased greatly. Products are available for walling of low- to medium-strength requirement. Aerated blocks are reasonably frost-resistant, though they would normally be protected from the weather, if used externally, by cladding or rendering.

No-fines concrete

As the name implies, this concrete is produced from cement and coarse aggregate only, the latter usually being graded between 20 and 10 mm. The aggregate may be ordinary dense aggregate, which would result in the concrete having a density in the region of 1800 kg/m^3, or lightweight aggregate, resulting in a concrete density as low as 500 kg/m^3. Dense aggregates would produce strengths of up to 15 N/mm^2, according to cement content, while the ceiling strength for lightweight aggregates would be approximately 10 N/mm^2.

In producing no-fines concrete, great care should be taken to obtain the correct water content, since a low value would result in uneven distribution of cement, while a high value would cause settlement of the cement paste fraction to the base of the concrete. Only minimal vibration should be used for compaction.

Shrinkage of no-fines concrete is low, since the cement paste is discontinuous, though shrinkage occurs quickly due to the high air permeability of the material. No-fines concrete is resistant to frost, though rendering of exposed surfaces is normally carried out to prevent water penetration.

No-fines concrete was formerly used as an in-situ walling material for low-rise dwellings. It is also manufactured in the form of lightweight precast blocks.

3.18 High-strength concrete

The term 'high-strength concrete' (see also section 2.5) generally refers to concretes having 28-day compressive strength over 60 N/mm^2. Such concretes may offer the following advantages:

- Increased hardening rates leading to reduced construction times.
- Smaller section sizes for a given load, leading to economies of space.
- Greater stiffness, hardness, chemical resistance and durability.

The DoE method of mix design can be used to produce high-strength concrete by the use of one or more of the following modifications:

1. Use superplasticisers to obtain water/cement ratios of 0.3 or less.
2. Produce rich mixes of very low workability. It should be noted that such mixes will require pressure/vibration for compaction and therefore have limited site application.

Their use would also be confined to situations where increased shrinkage is not a problem – for example, in precast units.
3. Use vacuum de-watering followed by power trowelling (for floors).
4. High-strength precast units can be made by autoclaving mixes containing pozzolanas.

To obtain concrete of strength up to 100 N/mm^2 the use of high-strength aggregate such as crushed granite may be essential since low-strength aggregates, or those having smooth surfaces, will tend to fail or become debonded under stress. Such concretes are readily obtainable either site mixed or ready mixed. Production of strengths of over 100 N/mm^2 is likely to be confined to precasting works that have suitable facilities.

One of the factors to be considered when employing very high-strength concretes is the applicability of normal structural design principles. It is found that very strong concretes do not have proportionately higher stiffness (elastic modulus) and this implies that strains/deflections in service will be higher than with conventional concrete. The effects of this on the behaviour of reinforcing steel, as well as on the structure as a whole, should be considered as part of the design process.

3.19 Production of high-quality surface finishes

The smooth off-white finish of newly produced concrete is usually attractive, often giving clean lines and a pleasing appearance to the structure as a whole. It is unfortunate that in many cases the appearance is not retained in the long term and older buildings often take on a 'drab' appearance even if the concrete remains in good condition. Some of the reasons for these changes are:

1. The surface of even good-quality concrete can be quite permeable. The production of a good-quality surface involves the presence of a layer of fine material conforming to the formwork surface. Such layers are usually of higher permeability than the core of the concrete. Hence they can readily absorb dirt, especially if not exposed to the cleaning action of rain.
2. More coarsely textured surfaces provide sites for dirt to be trapped within their texture (Fig. 3.24).
3. Surfaces quite commonly contain small 'defects' such as blow holes resulting from entrapped air. (These can be minimised by correct choice of release agent for the formwork material being used.)
4. In general, the concrete surface cannot be inspected until it is hard. Therefore experienced personnel are required to judge whether the concrete has been correctly and fully compacted. Honeycombing in areas where coarse aggregate is insufficiently enclosed in mortar is quite common.
5. Joints in formwork may leak and lead to grout loss.
6. In-situ concrete must be provided with construction joints. Since it can be almost impossible to provide an invisible joint, they are best made a feature of the design, so that the joint profile hides imperfections in the joint detail.

Preferred finishes have, to some degree, been a matter of fashion at the time of construction. For example, in the 1970s, sawn board finishes (Fig. 3.24) were often favoured. However, they were very expensive since they required solid sawn timber for

Fig. 3.24 *Typical weathered sawn board finish. The holes are for fixing purposes.*

formwork, and labour costs in fixing were also high. The final finish also tends to attract dirt, producing a predominantly dull effect, especially if used in large areas.

Current practice tends to avoid large areas of plain concrete; instead the finish can be textured, for example:

- Bush hammering to remove the permeable laitance layer and to provide a partially exposed aggregate surface texture.
- Application of a surface retarder. The surface remains soft and can be brushed off to reveal aggregate.
- Use of textured formwork to provide an interesting concrete texture.
- Use of aggregate transfer techniques so that the surface appearance is dominated by the aggregate itself.

Figure 3.25 shows a close view of a building, now 10 years old, in which several techniques have been combined.

- The precast, smooth-finish, light-coloured sill units have weathered well, except in a recess where dirt has gathered.
- The exposed aggregate precast concrete blocks (comprising an integral insulating backing) have shed the dirt very effectively.
- The bush-hammered in-situ concrete (buttress) has a rust stain due to corroding reinforcement on one arris, and has darkened due to damp near ground level.
- The exposed aggregate paving is affected by moss growth (little exposure to the sun on the north side of the building).

These points illustrate the factors to be considered if designing concrete to have long-lasting, high-quality appearance. The need for the very highest quality procedures at each stage of production, therefore, cannot be over emphasised.

Fig. 3.25 *Weathering of various concrete finishes in a 10-year-old building.*

3.20 Recycling concrete

In terms of volume, concrete is now one of the most widely used materials in the construction industry and, consequently, its treatment at the point of demolition is attracting much interest. With landfill taxes likely to rise sharply, some form of recycling is increasingly making economic as well as environmental sense. The main factors affecting recycling are:

1. The physical form of concrete can make handling and recycling difficult, especially if reinforced. Heavy machinery is essential for separating steel and for crushing the concrete. For this reason, recycling is restricted to situations where such plant is accessible. The initial stages of recycling concrete containing steel involve breaking up in-situ frames and then reducing columns and beams to a more manageable size by large scissor-action machines which cut through the concrete and steel. These produce material which is easier to handle or transport. Mobile crushers are available but transport costs restrict their applications to situations in which large quantities of concrete are involved.
2. Pure crushed concrete (for example, from highways works) may occasionally be available but demolition waste is normally contaminated by brick and other waste, which may not be acceptable for some applications.
3. Unless extensive stockpiling facilities are available, demand for the recyclate must be roughly matched to the supply of concrete waste.

Possible applications

Road sub-bases

Pure crushed concrete may well satisfy requirements for use in sub-bases (see Section 11.4) but compositional limits of such specifications are not usually satisfied by demolition waste. However, there is evidence that acceptable behaviour can be obtained for sub-base applications and there is perhaps the need for additional flexibility to be built in (for example, by concentrating on performance rather than composition) to permit maximum reuse in lower performance applications such as housing estate roads.

One Dutch specification that has been used successfully for some years requires at least 50 per cent of the material to be crushed concrete or equivalent, but up to 50 per cent to be other crushed masonry of density at least 1600 kg/m³ (the latter density range includes crushed brick). There are also particle-grading requirements and limits to organic materials such as plastics (1 per cent by mass) and perishable materials such as paper (0.1 per cent by mass).

The specification can be satisfied relatively easily by construction waste as follows:

Separate large timber pieces:	hand
Primary size reduction/screening:	machine
Separation of steel:	electromagnet
Separation of non-ferrous metal:	hand
Separation of wood, plastic, paper:	machine or hand
Secondary crushing/screening:	machine

The labour and plant requirements for the process are evident and it is therefore best carried out on a relatively large scale.

Recycled concrete

1. When used as a source of aggregate for concrete, crushed concrete has in general lower strength and higher water absorption characteristics than primary aggregates. This makes mix design more difficult.
2. Since the proportion of cement paste in recycled concrete is much larger than in the original material, shrinkage and creep will be increased.
3. Perhaps the largest problem for reuse as concrete is that large variations in quality can be expected. This, in general, rules it out for structural concrete. Such variations can, however, be minimised by the use of several stockpiles which are then blended to average out composition over longer time periods.

The above suggests that unless uniform sources are available, the use of construction waste should focus upon non-structural applications and those without demanding exposure requirements. There are, however, many such applications in low-rise construction. The operations referred to in sub-base applications could form a basis for suitable aggregate with the further requirement that the material is washed to remove very fine or other unsuitable components. Figure 3.26 shows a sample of concrete obtained from construction waste, processed as above. It includes brick waste. The surface was treated with a surface retarder to reveal an attractive range of colours from the different materials. The product is noticeably permeable but may be suitable for non-exposed applications.

Fig. 3.26 *Recycled concrete slab made from construction waste sorted and processed as above. There is a range of particle shapes present, including crushed brick.*

EXPERIMENTS

(See the British/European standards listed at the end of the chapter for full details and specifications.)

Experiment 3.1 Grading of aggregates (BS 812: section 103.1; BS 882)

Note that the term 'sand' rather than 'fine aggregate' is used in the experiment to signify material passing a 5 mm sieve in accordance with BS 882 (1992).

Apparatus
BS sieves; weighing balance accurate to 1 g.

Sample preparation
A main sample is first obtained from the stockpile by the procedure outlined in BS 812: Part 101. The essential requirement is that this sample should consist of at least 10 increments, obtained from different parts of the stockpile and mixed thoroughly. The main sample should be of not less than 13 kg for sand and not less than 25 kg for aggregates of nominal size between 5 and 28 mm.

The main sample is reduced in size to a quantity suitable for sieving, using a riffle box (Fig. 3.27), or by quartering. It is recommended that division be continued until the minimum mass in excess of the requirements of Table 3.15 is obtained. Larger masses, particularly of sand, tend to lead to blinding of sieve apertures. Note, however, that shaking a divided sample from a scoop to obtain the exact minimum mass of Table 3.15 is incorrect – it leads to particle segregation. The prepared sample must then be dried either in an oven or by leaving for some time on a flat sheet.

Procedure
Select the sieves appropriate to the nominal size of material being used by reference to Tables 3.17 and 3.18. Examples are:

Fig. 3.27 *Riffle box for reduction of aggregate sample size. Three receiving boxes are normally supplied.*

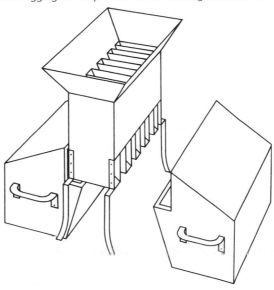

Table 3.15 *Minimum mass of sample for sieve analysis*	
Maximum size of material present (mm)	Minimum mass of sample to be taken for sieving (kg)
50	35
40	15
20	5
14	1.0
10	0.5
5	0.2
3	0.2
< 3	0.1

- **20 mm graded coarse aggregate or 20 mm single size coarse aggregate:**
  37.5, 20, 14, 10, 5 mm
- **10 mm single size coarse aggregate:**
  14, 10, 5, 2.36 mm
- **sand:**
  10, 5, 2.36, 1.18 mm; 600, 300, 150 μm

Table 3.16 *Maximum amount to be retained on each sieve*

To avoid 'blinding' of apertures

BS test sieve nominal aperture size (mm)	Maximum mass (kg)		BS test sieve nominal aperture size		Maximum mass (g)
	450 mm diameter sieves	300 mm diameter sieves	(mm)	(µm)	
50.0	14	5	2.36	–	200
37.5	10	4	1.18	–	125
20.0	6	2.5	–	600	100
14.0	4	2.0	–	425	80
10.0	3	1.5	–	300	65
5.0	1.5	0.75	–	150	50

Table 3.17 *Grading limits for sand*

Sieve size	Percentage passing BS sieves		
	F	M	C
10 mm	100	100	100
5	89–100	89–100	89–100
2.36	80–100	65–100	60–100
1.18	70–100	45–100	30–90
600 µm	55–100	25–80	15–54
300	5–70	5–48	5–40
150	0–15	0–15	0–15

Shake the material through each sieve individually into a collecting tray starting with the largest, using a varied motion, backwards and forwards, clockwise and anticlockwise and with frequent jarring. (Alternatively, sieves may be assembled into a sieve shaker.) In order to prevent blinding of sieve apertures, the quantity of material retained on sieves should not exceed the values given in Table 3.16. Smaller sieves often give problems in this respect and one solution is to use additional sieves – for example, a 425 µm sieve prior to the 300 µm sieve. The quantities retained on the sieves are then added. Weigh the quantities retained on each sieve and check that the total is approximately equal to the original sample size. If it is not, repeat the procedure with a new sample.

Table 3.18 *Grading limits for coarse aggregates*

	Percentage passing BS sieve sizes							
BS sieve (mm)	*Nominal size of graded aggregate (mm)*			*Nominal size of single-size aggregate (mm)*				
	40–5	*20–5*	*14–5*	*40*	*20*	*14*	*10*	*5*
50.0	100			100				
37.5	90–100	100		85–100	100			
20.0	35–70	90–100	100	0–25	85–100	100		
14.0	25–55	40–80	90–100	–	0–70	85–100	100	
10.0	10–40	30–60	50–85	0–5	0–25	0–50	85–100	100
5.0	0–5	0–10	0–10	–	0–5	0–10	0–25	45–100
2.36	–	–	–	–	–	–	0–5	0–30

Complete a table similar to Table 3.6 using appropriate sieve sizes. Plot the percentage passing each sieve using percentage passing as the *y* axis and sieve sizes increasing equally spaced as the *x* axis. Join each point by a straight line. Classify sands into zones C, M or F by reference to Table 3.17. Note that sands may comply with more than one grading; if they do, record the fact and identify which zone fits best. Table 3.18 shows limits for coarse aggregates.

Note that the figure of 15 per cent for the 150 μm sieve is increased to 20 per cent for crushed rock sand except when they are used for heavy duty floors.

Experiment 3.2 Determination of bulking of sand

Apparatus
250 ml measuring cylinder; 5 ml measuring cylinder; large beaker; wide funnel; spatula; dry sand.

Procedure
Weigh out 250 g of sand and transfer to the 250 ml measuring cylinder using the funnel. Tap gently to level the sand and note the volume. Transfer the sand to the beaker and add 5 ml (2 per cent) of water. Mix and again transfer to the measuring cylinder, noting the new volume after tapping. Repeat the procedure using 2 per cent increments of water until free-moisture begins to separate from the sand – usually at about 20 per cent moisture content. Note that to achieve consistent results, a standardised tapping (levelling technique) is necessary.

Results
Measure each volume as a percentage of the original dry volume and plot graphically against moisture content. (Note that even dry fine sand contains substantial quantities of air.)

Experiment 3.3 Determination of the volume of silt and clay (fines) in a sample of sand using the field-settling test

Note
1. The term 'fines' is used here to denote silt, clay or dust in accordance with BS 882 (1992).
2. The field-settling test is no longer a BS test.

Apparatus
A 250 ml measuring cylinder; common salt; sample of sand (not crushed stone sand).

Procedure
Prepare a 1 per cent solution of sodium chloride (common salt) in water. Two tea-spoonfuls of salt to 1 litre of water will suffice. Pour 50 ml of this solution into the 250 ml measuring cylinder. Gradually add the sand until the volume of the sand is 100 ml. Make the volume of solution up to 150 ml by adding more solution. Shake vigorously until the clay-type particles are dispersed and place on a level bench and tap gently until the surface of the sand is level. After three hours, measure the volume of silt above the sand/silt interface and express this as a percentage of the volume of sand (Fig. 3.7). Refer to the text of this chapter (page 57) for the acceptable limits.

Experiment 3.4 Measurement of the moisture content of aggregates using the siphon can

Apparatus
Siphon can (Fig. 3.28); weighing balance of 3 kg capacity, capable of weighing to 1 g accuracy; 500 ml measuring cylinder.

Procedure
After ensuring that the siphon can is clean and empty, close both siphon tubes and fill until the water is above the highest point on the upper tube. Open the upper tube and discharge the water to waste. Open the lower tube and collect the water in the measuring cylinder. The volume V ml is the calibration volume for the can. Repeat and obtain an average value for V.

To find the moisture content of an aggregate, fill the siphon can with water and discharge using the lower siphon tube (this condition is obtained after the calibration test). Close the lower siphon tube. Weigh 2 kg of dry aggregate to 1 g accuracy and carefully add to the siphon can, stirring to remove air bubbles. Allow the stirring rod to drain and remove it. Open the upper siphon tube and collect the water in the measuring cylinder (volume v_b). Open the lower siphon tube and, after draining, close both tubes. Repeat using 2 kg of the damp aggregate whose moisture content is required, obtaining the volume v_w ml.

Results
Calculate the moisture content of the aggregate using the equations:

Fig. 3.28 *Siphon can for moisture content determination.*

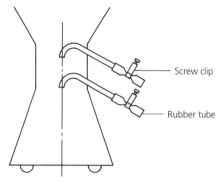

Screw clip

Rubber tube

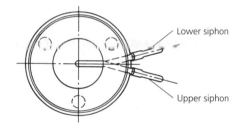

Lower siphon

Upper siphon

$$\text{moisture content (\% by wet mass)} = \frac{v_w - v_b}{1000 - V - v_b} \times 100$$

$$\text{moisture content (\% by dry mass)} = \frac{v_w - v_b}{1000 - V - v_w} \times 100$$

Note

These equations are based on the assumption that no water is absorbed by the aggregates, hence that the 'dry' sample is in the saturated surface-dry condition and that free-water is present in the damp sample. These should apply in most cases but, where aggregates absorb water, the procedure given in BS 812 should be followed.

Experiment 3.5 Measurement of the moisture content of sand using the calcium carbide method

Apparatus
'Speedy' moisture meter; calcium carbide.

Procedure
Make sure the apparatus is clean and dry before commencing the test. Mix the main sand sample and then immediately measure out the sand sample for testing, using the scoop provided. Place this sample in the cap of the apparatus. Use the scoop to place the correct quantity of calcium carbide in the body of the instrument. Bring both parts

together, keeping them horizontal to avoid mixing of the samples and tighten the cap, ensuring that the clamp screw is correctly located. With the dial facing downwards, shake the instrument for three seconds and then stand, dial facing upwards. Tap to locate material in the cap and allow to stand for one minute. Repeat the shaking operation. Finally, read the instrument with the dial vertical. The reading gives the moisture content, based on wet weight directly. Empty the instrument out of doors and repeat with another sample, obtaining the average reading.

Experiment 3.6   Determination of the slump of a concrete mix

Apparatus
Slump cone; hard, level vibration-free, non-absorbent base; tamping rod; rule trowel.

Procedure
Ideally the concrete should be sampled from a large batch and prepared in accordance with BS 1881: Part 101. Alternatively, 8 litres of concrete to the specification on page 78 may be produced, assuming aggregates of this specification are available.

If neither of the above is convenient, a sample of about 8 litres may be made up using masses as follows: 3 kg cement, 5 kg sand, 8 kg coarse aggregate.

The materials are mixed dry and sufficient water (about 1.5 litres) is added to produce a workable concrete. Place the clean slump cone on the base and hold firmly in position throughout the filling operation. Load concrete gently into the cone until a depth of about 100 mm is obtained. Rod 25 times with the rounded end of the tamping rod. Add another 100 mm layer and repeat, rodding through to the underlying layer. Repeat until three compacted layers are obtained and strike the top layer off level with the trowel. While still holding the cone, clean away excess concrete from around the base, then remove the slump cone by lifting carefully in the vertical direction. Invert the cone, place next to the concrete and, using the tamping rod as a reference height, measure the slump (to the highest part of the slumped concrete) (Fig. 3.8). Record the slump in millimetres and the nature of the slump – true, shear or collapsed.

Experiment 3.7   Determination of the workability of concrete using the compacting factor apparatus

Apparatus
Compacting factor apparatus (Fig. 3.9); 2 metal floats; vibrating table or 25 mm square tamping rod for compaction; weighing balance capable of up to 25 kg to 0.1 g accuracy; trowel; slump test tamping rod.

Procedure
Either obtain a sample of concrete or prepare a sample, as described in Experiment 3.6. Very wet mixes should not be used.

Weigh the collection cylinder, place in position and cover with the two metal floats. Close both trap doors. Load the concrete gently into the upper hopper until it is well heaped. Open the upper trap door and allow the concrete to fall through into the lower hopper. If the concrete 'hangs up' in the hopper, it may be rodded through gently with the slump test tamping rod. Remove the floats from the receiving cylinder and open the lower trap door, allowing the concrete to fill the cylinder. Cut away excess concrete by sliding the metal floats across the top of the receiving cylinder. Wipe the outside of the cylinder clean and weigh to the nearest 0.1 kg. Subtract the weight of the cylinder to obtain the weight of partially compacted concrete. Empty the cylinder and refill in 50 mm layers, tamping each layer or compacting using a vibrating table to obtain full compaction. Weigh again and hence obtain the mass of fully compacted concrete. The compacting factor is then the ratio of the two masses:

$$\frac{\text{mass of partially compacted concrete}}{\text{mass of fully compacted concrete}}$$

Experiment 3.8 Measurement of the workability of concrete using the Vebe consistometer

Apparatus
Vebe consistometer and vibrating table (Fig. 3.10); trowel; stopwatch.

Procedure
Either obtain a suitable sample of concrete or prepare a sample, as described in Experiment 3.6.

This method is most suited to dry mixes – the concrete should be barely plastic with very little or no slump. Fix the slump cone and outer container in position and attach the filling cone. Compact the concrete into the cone as in the slump test (Experiment 3.6). Remove the filling cone and the slump cone and swing the Perspex disc into position, making sure the disc is free to move vertically in its guide. Rest the disc gently on the concrete. Switch on the vibrator and measure the time taken for the disc to sink, so that the concrete wets it uniformly all round. The time taken in seconds is then the consistency of the concrete in 'Vebe degrees'. It will be clear that stiffer mixes produce higher results.

Experiment 3.9 Measurement of the compressive strength of concrete using the cube test

It is suggested that the greatest value will be obtained from this experiment if the concrete for cube tests is produced as a series of trial mixes which will enable the effect of proportions on strength to be investigated.

Apparatus
Compression machine of 1000 kN or greater capacity; 100 or 150 mm cube moulds, preferably steel; vibrating table or tamping rod for compaction.

Procedure
First ensure that all the moulds are in good condition and that all clamps or bolts are tight. Coat all internal surfaces with a thin film of mould oil. Mix the concrete sample thoroughly and place in the cube moulds to a depth of approximately 50 mm. At least five cubes per concrete sample are recommended. Compact this layer using a vibrating table or by hand and add another 50 mm layer, repeating the procedure. When the

Fig. 3.29 *Modes of failure of concrete cubes: (a) normal; (b) abnormal – tensile cracks marked T are indicative of incorrect testing or faulty machine.*

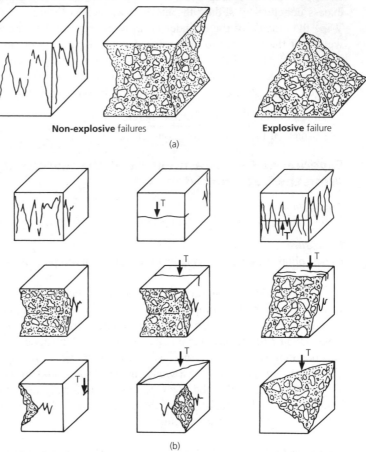

Non-explosive failures   **Explosive** failure

(a)

(b)

moulds are full, trowel smooth and cover with an impermeable mat. Cubes must be identified, the most effective way initially being to use numbers painted on the moulds with mix and batch details being referenced.

Leave undisturbed for 16–24 hours at a temperature of 20 ± 2 °C and then demould, identifying each cube with waterproof ink. Transfer to a curing tank containing clean water and cure at 20 ± 1 °C until testing.

At the required age (usually 7 or 28 days), remove the cubes from the tank and wipe away any excess water or grit. Place cubes, trowelled face sideways, centrally in the testing machine and load at the BS rate until failure. Note the maximum load and the mode of failure, which should be one of those indicated in Fig. 3.29(a). If there are horizontal cracks in the concrete, a tensile failure has occurred and this must be reported. Calculate the stress in N/mm^2 by dividing the load in N by the cross-sectional area in mm^2.

Suitability of the testing machine

Some machines in use have been found to be unsatisfactory for cube testing. Machines should comply with BS 1881: Part 115 and periodic checks are necessary to ascertain that the upper ball seating works properly and that bearing surfaces are not excessively worn. The ball seating forms part of a sphere with its centre at the face of the upper platen. It is designed to rotate during initial loading to align accurately with the upper face of each concrete cube. Thereafter it should lock by friction in order to avoid premature crushing of any weak portion of the concrete cube (for example, the trowelled face). Where failures of the form of Fig. 3.29(b) occur, they may indicate problems with the ball seating. A good way of checking the machine generally is to use the 'reference test service' operated by a number of organisations with suitable machines. Pairs of 'identical' cubes are made and tested on each machine to reveal shortcomings in the machine in question; if the machine is in good condition, test results should be very similar.

Experiment 3.10 The effect of mix proportions on the properties of fresh and hardened concrete

Note

A comprehensive investigation into the effect of mix proportions on the properties of concrete will require the production of a considerable number of mixes. Where time permits, a series of nine mixes might be made, each mix being designed by the DoE method of mix design. Table 3.19 indicates nine such mixes that might be produced. Where time is limited, the study might be confined to three mixes, for example, numbers 1, 5 and 9. (These three mixes should help students to overcome the common fallacy that dry mixes are always stronger.)

Apparatus

As for Experiments 3.6 and 3.9.

Procedure

Select a number of mixes from Table 3.19 and produce batch quantities by use of the form of Fig. 3.12, using the same size, type and size or aggregate in each case. (Produce a target mean strength on the basis that no previous results are available.) For the purpose of the slump test, 8 litres of each mix will be required.

Table 3.19 *Suggested trial mixes*

28-day characteristic strength (N/mm²)	Slump (mm)		
	0–30	30–60	60–180
15	1	2	3
25	4	5	6
40	7	8	9

Measure the free-moisture contents of the aggregates and, for each mix, correct the batch quantities as described on page 59.

For each concrete mix proceed as follows: Mix the materials dry and add a quantity of water somewhat less than the required batch quantity. Try to estimate the likely slump and continue to add water until what is judged to be the correct slump is obtained. Carry out the slump test. Note that workability tests should, if possible, be carried out at a standard time (usually 6 minutes) after adding the mixing water. If the workability is outside the acceptable range, another mix should, strictly speaking, be produced.

The cohesiveness of the concrete should also be measured – produce a small heap of the concrete and trowel to a thickness of about 50 mm. A mix of normal cohesion should trowel to a smooth finish fairly easily. Mixes of high or low cohesion will be very easy or very difficult, respectively, to trowel to a smooth finish.

Make up at least five cubes from the batch, curing and testing as described in Experiment 3.9. All results should be recorded and compared with the original specification. The variability within each set of five cubes should also be measured by calculation of the standard deviation and comparison with values indicated on page 82. (Note that five results are not really sufficient for a full analysis and that laboratory-produced concrete should have a relatively low standard deviation.) The cost of the concrete varies directly with the cement content; note which mix is most expensive on this basis.

SELF-ASSESSMENT QUESTIONS

1. List the four chief components in Portland cement. State which of these:
   (a) gives off most heat during hydration;
   (b) is present in the largest quantity of OPC;
   (c) is vulnerable to sulphates;
   (d) is responsible for the grey colour of OPC.

2. State what is meant by a hydraulic cement.

3. Indicate two ways in which the manufacturer of a cement could increase its rate of hardening using given raw materials. Give typical applications of rapid-hardening cements.

4. State the cause of sulphate attack in concrete. Indicate two ways in which the likelihood of this attack can be reduced.

5. Name the two principal compounds produced on hydration of ordinary Portland cement. Give one characteristic property of reinforced concrete which depends on each.

6. Explain the purpose of the following tests for cement:
   (a) setting time;
   (b) soundness;
   (c) mortar cube strength.

7. Explain the factors which affect the choice of the maximum size of aggregate for concrete and indicate what maximum size of aggregate would be suitable for the following purposes:
   (a) a normally reinforced concrete column of section 600 mm × 400 mm;
   (b) a 40 mm floor screed;
   (c) a concrete road base of thickness 200 mm.

8. A sieve analysis of 250 g of sand gives the following results:

Sieve size	Mass retained
10 mm	0
5	5
2.36	31
1.18	38
600 mm	38
300	79
150	51
Passing 150 μm	8

   Classify the aggregate by means of Table 3.17.

9. Explain what is meant by 'bulking' of aggregates. Indicate why the extent of bulking of sand depends on moisture content.

10. A sample of damp aggregate weighing 2.35 kg is dried by hair dryer until it just reaches the free-running (saturated surface dry) condition. It is then found to weigh 2.24 kg. After drying in the oven at 110 °C to constant mass, it is found to weigh 2.15 kg. Based on dry mass, calculate:
    (a) the free-water content;
    (b) the total water content.

11. Compare the following tests for the workability of concrete:
    (a) slump test;
    (b) compacting factor test;
    (c) Vebe test.

    A sample of concrete was divided into three and results for the slump, compacting factor test and Vebe tests obtained. They were as follows:

    slump: 5 mm
    compacting factor: 0.93
    Vebe time: 18 s

    Assuming that only one result was in error, which result is incorrect? (Refer to Table 3.8.)

12. Explain what is meant by the term 'cohesion' in relation to fresh concrete. Indicate the problems that might arise in concretes of
    (a) very low cohesion;
    (b) very high cohesion.

13. Use Table 3.11 and Fig. 3.14 to determine the water/cement ratios required in mixes to the following specifications:

	Target mean strength at stated age (N/mm)^2		
Cement/aggregate	3 days	7 days	28 days
1. OPC/uncrushed	25	22	40
2. OPC/crushed	20	40	60
3. RHPC/crushed	30	35	60

14. Use the DoE method of mix design to produce batch quantities for 50 litres of concrete to the following specification:

Cement:	Ordinary Portland
Aggregates:	Sand – uncrushed; relative density 2.6; 59 per cent passes the 600 μm sieve
	Coarse – 20 mm, uncrushed; relative density 2.6.
Compressive strength:	35 N/mm^2 at 28 days (5 per cent defectives), no previous results available
Workability:	30–60 mm slump

15. Use the DoE method of mix design to produce batch quantities for 10 litres of concrete to the following specification:

Cement:	Rapid-hardening Portland
Aggregates:	Sand – uncrushed; relative density 2.7; 75 per cent passes the 600 μm sieve
	Coarse uncrushed, relative density 2.7; maximum size 40 mm
Min. cement content:	350 kg/m^3 concrete
Compressive strength:	35 N/mm^2 at 28 days (5 per cent defectives allowed)
Standard deviation:	5 N/mm^2
Workability:	10–30 mm slump

16. Concrete is required to be produced with a characteristic strength of 25 N/mm^2 with 2 per cent failures allowed. The first 20 cube results are as follows:

24.2	23.9	28.1	20.2	29.4
32.5	19.8	24.2	26.5	23.2
30.5	27.5	24.1	26.5	29.1
20.1	24.5	25.1	22.9	21.2

Calculate the standard deviation using the formula given on page 81 and hence indicate whether the required characteristic strength is being reached.

17. Concrete for a particular contract is to be produced with a target mean strength of 33 N/mm^2 and the assumed standard deviation is 4.5 N/mm^2. The first 50 results are as follows (in N/mm^2):

26.2	28.0	34.8	39.0	40.0	28.2	30.0	38.6	37.0	33.0
34.6	33.6	35.0	29.6	28.8	30.0	26.0	24.4	29.2	36.0
40.0	30.2	33.4	36.8	34.9	26.0	26.8	30.6	37.6	31.6
30.2	28.4	37.2	35.6	30.6	40.6	37.2	30.4	33.4	34.8
38.6	33.8	33.6	28.8	32.8	34.8	39.0	28.4	34.6	32.8

Draw a control chart of the type shown in Fig. 3.19, including the target mean strength and upper and lower control lines for 5 and 10 per cent of *individual* results. Hence suggest whether the characteristic strength requirement of 25 N/mm^2 with 5 permissible failures is being satisfied.

18. Concrete is required to have a characteristic strength of 30 N/mm^2 at 28 days with 5 per cent permissible failures. Calculate the water/cement ratio that will be needed if OPC and uncrushed aggregates are used. Assume a standard deviation of 5.5 N/mm^2.

19. Indicate three alternative reasons for the use in concrete of
   (a) plasticisers;
   (b) retarders.

20. Define 'drying shrinkage' in the context of concrete and indicate typical shrinkage strains that might be obtained. Compare these with the tensile strain capacity of mature concrete. Identify one situation in which allowance for shrinkage must be made.

21. For each of the following identify the root cause of the problem in types of concrete and situations likely to be affected, and state the mix parameters that should be used to avoid each problem:
   (a) sulphate attack;
   (b) frost attack;
   (c) alkali-silica reaction.

22. Describe the chief factors that should be considered when producing long-lasting high-quality concrete surfaces.

REFERENCE

Design of normal concrete mixes. Department of the Environment, HMSO, 1988, 1997.

STANDARDS

BS 12: 1996: *Portland cement*
BS 146: 1996: *Portland blast furnace cements*
BS 812: *Testing aggregates*
   Part 101: 1984: *Guide to sampling and testing aggregates*
   Part 102: 1989: *Methods for sampling*
   Part 103: *Method for determination of particle size distribution*
      Section 103.1: 1985: *Sieve tests*
      Section 103.2: 1989: *Sedimentation test*
   Part 109: 1990: *Methods of determination of moisture content*

BS 882: 1992: *Specification for aggregate from natural sources for concrete*
BS 1370: 1979: *Low-heat Portland cement*
BS 1881: *Testing concrete.*
    Part 101: 1983: *Methods of sampling fresh concrete on site*
    Part 102: 1983: *Method for determination of slump*
    Part 103: 1993: *Method of determination of compacting factor*
    Part 104: 1983: *Method of determination of Vebe time*
    Part 108: 1983: *Method of making test cubes from fresh concrete*
    Part 111: 1983: *Method of curing of test specimens (20 °C method)*
    Part 115: 1986: *Specification for compression test machines for concrete*
    Part 116: 1983: *Method of determination of compressive strength of concrete cubes*
BS 4027: 1996, *Sulphate resisting Portland cement*
BS 5075: Part 1: 1982: *Accelerating admixtures, retarding admixtures and water-reducing admixtures*
    Part 2: 1982: *Air-entraining admixtures*
    Part 3: 1985: *Superplasticising admixtures*
BS 5328: *Concrete*
    Part 1: 1997: *Guide to specifying concrete*
    Part 2: 1997: *Methods for specifying concrete mixes*
BS 6588: 1996: *Portland pulverised-fuel ash cements*
BS 6610: 1996: *Pozzolanic pulverised-fuel ash cement*
BS 8110: *Structural use of concrete*
    Part 1: 1997: *Code of practice for design and construction*
BS EN 196: *Methods of testing cement*
    EN 196–1: 1995: *Determination of strength*
    EN 196–3: 1995: *Determination of setting time and soundness*
BS EN 197: *Cement. Composition, specifications and conformity criteria*
    EN 197–1: 1995: *Common cements*
DD ENV 206: 1992: *Concrete. Performance, production, placing and compliance criteria*

4 Masonry

Chapter summary

- Definition of terms.
- Clay, calcium silicate and concrete bricks; their properties and uses.
- Durability of brickwork.
- Building blocks
- Natural stone masonry.
- Masonry mortars.
- Renderings.
- Porosity, capillarity, permeability and dampness.
- Recycling masonry.

4.1 Introduction

Masonry remains a key component of most buildings. Even steel-framed and timber-framed buildings often have a masonry cladding. An understanding of the attractions and potential problems of masonry is therefore most important.

The great majority of masonry walling is produced from preformed units, bonded by a mortar. These may be subdivided into

- **Bricks**, which are easily handled with one hand.
- **Blocks**, which are larger units, generally requiring both hands.

Each of these forms has advantages – for example, the smaller size of bricks may be a visual advantage in smaller buildings such as house construction, and facing clay units, which are often of very attractive appearance, are difficult to produce in larger sizes. Conversely, use of blocks permits rapid and economical construction and a great range of strengths, finishes, densities and thermal properties can be obtained.

Environmental considerations are presented in Table 4.1.

Table 4.1 *Environmental considerations: masonry*

Consideration	Assessment
Raw material availability	Most materials are abundant
Extraction	Care is required to avoid damage to the landscape, particularly after extraction has ceased
Energy used in manufacture	Both cements and clay brick manufacture involve high temperatures
Health/safety hazards	Few specific hazards; inhalation of any dust to be avoided
Recyclability	Bricks and natural stone laid in lime mortars may be reused; use of other forms mainly limited to crushing/hard-core; roofing slates readily reuseable

Fig. 4.1 *Relationship between the dimensions of a standard metric brick.*

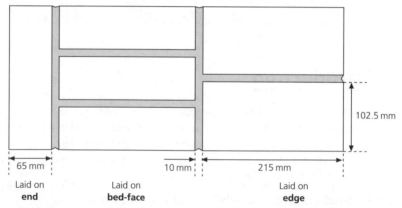

65 mm	10 mm	215 mm	102.5 mm
Laid on **end**	Laid on **bed-face**	Laid on **edge**	

4.2 Bricks

Bricks may broadly be described as building units which are easily handled with one hand. By far the most widely used size at present is the single standard metric brick of actual size 215 × 102.5 × 65 mm. The relationship between these dimensions is illustrated in Fig. 4.1. This gives great versatility to the use of standard metric bricks.

It should be noted that the standard metric brick is of smaller length and depth than imperial bricks used extensively prior to the 1970s. Careful consideration should therefore be given when alteration of, or extension to, older brick buildings is planned.

For the past 100 years, clay bricks have dominated the UK market as building units, but more recently other brick types and concrete or clay blocks have provided strong competition, especially since thermal insulation regulations were tightened and light-weight forms became available.

Indentations and perforations in bricks

Indentations (frogs) and perforations (cylindrical holes passing through the thickness of the brick) may be provided for one or more of the following reasons:

- They assist in forming a strong bond between each brick and the wall as a whole.
- They reduce the material cost and hence the overall cost of the brick without serious in-situ strength loss.
- In the case of clay bricks they reduce the effective thickness of the brick and hence reduce firing time since heat penetrates the clay more easily.

For greatest strength, bricks with a single frog should be laid frog-up, since this ensures that the frog is filled with mortar. Perforated bricks generally contain less than 25 per cent of voids and these behave, in a structural sense, as if they were solid.

Clay bricks (BS 3921)

In terms of availability, clay bricks continue to be a most important building unit, combining excellent durability (when they are selected and used correctly) with, in the case of facing bricks, lasting aesthetic properties. They are made by pressing a prepared clay sample into a mould, extracting the formed unit immediately and heat-ing it in order to sinter (partially vitrify) the clay. Many different types of brick may be produced, depending on the type of clay used, the moulding process and the firing process.

There are three basic subdivisions of type:

- **Common bricks**
  These are ordinary bricks which are not designed to provide good finished appear-ance or high strength. They are therefore in general the cheapest bricks available.
- **Facing bricks**
  These are designed to give attractive appearance, hence they are free from imperfec-tions such as cracks.
- **Engineering bricks**
  These are designed primarily for strength and durability. They are usually of high density and well fired.

Bricks may also be classified in other ways – for example, by their frost resistance or their soluble salt content, and many clay bricks are designated by their place of manu-facture, colour or surface texture, for example:

- **Fletton** – a common brick manufactured from Oxford clay, originally in Fletton, near Peterborough.

- **Staffordshire blue** – an engineering quality brick produced from clay which results in a characteristic blue colour.
- **Dorking stock** – the term 'stock', originally a piece of wood used in the moulding process, now denotes a brick characteristic of a certain region.

Manufacture

There are four basic stages in brick manufacture, though many of the operations are interdependent. A particular brick will follow through these stages in a way designed specifically to suit the raw material used and the requirements of the final product.

A key property of the clay in relation to brick manufacture is its moisture content at the time of processing.

A very dry clay has the potential to produce a very strong brick since, on firing, any moisture present evaporates leaving voids which in turn cause weakness. However, dry clays may have low plasticity and may be difficult to compact; therefore, intense compactive effort would be needed to remove air from the clay during the formation process. The water content may be increased after extraction to assist production but it is much more difficult to reduce the water content of a clay prior to moulding. (The situation is rather similar to that by which concretes are produced, a key factor then being the water/cement ratio and low values giving potentially the best performance but tending to require intense compactive effort.)

Clay preparation

After digging out, the clay is prepared by crushing and/or grinding and mixing until it is of a uniform consistency. Water may be added to increase plasticity (a process known as 'tempering') and in some cases chemicals may be added for specific purposes – for example, barium carbonate, which reacts with soluble salts producing an insoluble product and therefore reducing risk of problems such as efflorescence, crystallisation damage, or chemical attack of the mortar.

Moulding

A key requirement of all clay bricks is that, after moulding, the 'green' brick has sufficient rigidity to be demoulded immediately, otherwise the cost of producing the brick would be greatly increased. The moulding technique is designed to suit the moisture content of the clay, the following methods being described in order of increasing moisture content.

Semi-dry process
This process utilises a moisture content in the region of 10 per cent. The ground and screened material has a rather dry granular consistency which is still evident in the

fractured surfaces of the fired brick. The material is compacted into the mould, under high pressure, in up to four stages. After pressing, the faces of the brick may be sanded, or textured to produce a 'rustic' appearance. A most important product of this process is the 'Fletton', named after its place of origin near Peterborough. The clay contains some organic material which helps to fire the brick and therefore reduces the firing cost. The final product is so keenly priced that it can be transported over quite large distances and still be competitive with other more locally produced 'common' brick types. The plain Fletton is characterised by pink 'kiss' marks where the bricks were in contact in the kiln and has good suction properties for rendering. The low degree of vitrification leads to good cutting properties but also to poor durability in situations of high exposure – it generally needs the protection of a roof. Sandfaced products for face use should be handled with care; the surfaces and edges are prone to abrasion and are quite easily damaged.

Stiff plastic process

This utilises clays which are tempered to a moisture content of about 15 per cent. A stiff plastic consistency is obtained, the clay being extruded and then compacted into a mould under high pressure. The relatively low moisture content, combined with good compaction and thorough firing, produces a brick with low absorption and good strength properties. Many engineering bricks are made in this way, the clay for these often containing a relatively large quantity of iron oxide which helps to promote fusion during firing.

The wire-cut process

The clay is tempered to about 20 per cent moisture content and must be processed to form a homogeneous material. This is extruded to a size which allows for drying and firing shrinkage. Units are then cut to the correct thickness by tensioned wires. Perforated bricks are made in this way, the perforations being formed during the extrusion process. Wire-cut bricks are easily recognised by the perforations or 'drag marks', caused by the dragging of small clay particles under the wire. Modern firing processes lead to consistency of firing and a hard product with regular surface and dimensions.

Soft mud process

As the name suggests, this process utilises a clay, normally from shallow surface deposits, in a very soft condition, the moisture content being as high as 30 per cent. Breeze or town ash may be added to provide combustible material to assist firing or improve appearance. The clay is pressed into moulds which are sanded to prevent sticking. The green bricks are very soft and must be handled carefully prior to drying. Hand-made bricks are produced by a similar process, except that 'clots' of clay are thrown by hand into sanded moulds. This produces a characteristic surface texture (Fig. 4.2), which has great aesthetic appeal and is the reason for the continued production, on a limited scale, of hand-made bricks. The bricks are often subject to some distortion, especially if older firing methods such as clamps are used, but this can increase their aesthetic interest and appeal.

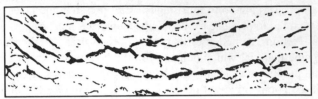

Drying

This must be carried out prior to firing when bricks are made from clay of relatively high moisture content. Drying enables such bricks to be stacked higher in the kiln without lower bricks becoming distorted by the weight of bricks above them. The process can often be carried out at reasonable cost by warm air from waste kiln heat. Drying also enables the firing temperature to be increased more rapidly without such problems as bloating, which may result when gases or vapour are trapped within the brick. Drying is carried out in chambers, the temperature being increased and the relative humidity progressively decreased as bricks lose moisture. The process normally takes a number of days, higher moisture content bricks requiring a longer period. Wire-cut bricks and those produced by the soft mud process must be dried prior to firing.

Firing

The object of firing is to cause localised melting (sintering) of the clay, which increases strength and decreases the soluble salt content without loss of shape of the clay unit. The main constituents of the clay – silica and alumina – do not melt, since their melting points are very high; they are merely fused together by the lower melting point minerals such as metallic oxides and lime. The main stages of firing are:

100 °C	evaporation of free-water
400 °C	burning of carbonaceous matter
900–1000 °C	sintering of clay.

The latter stages of firing may be assisted by fuels either present naturally in the clay or those added during processing – one of the factors contributing to the low cost of Fletton bricks is the presence of carbonaceous material, which is largely responsible for the heat supplied in the latter stages of firing.

Maintaining control of the temperature is most important in order to produce bricks of satisfactory strength and quality; in particular, too rapid firing will cause bloating and overburning of external layers, while too low a temperature seriously impairs the strength and durability. Stronger bricks, such as engineering bricks, are normally fired at higher temperatures. The main processes in firing are given below.

Clamps

Bricks are stacked in large special formations on a layer of breeze, though the bricks also contain some fuel. The breeze base is ignited and the fire spreads slowly through the stack, which contracts as the bricks shrink on firing. The process may take up to one

Fig. 4.3 *Continuous kiln showing sequence of zones and movement of air for drying purposes.*

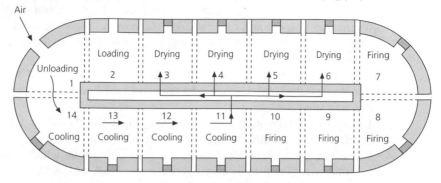

month to complete and the fired product may be very variable – some underburnt and overburnt bricks being obtained. After firing, the bricks are sorted and marketed for various applications. Well-fired bricks are extremely attractive with local colour variations caused by temperature differences and points where fuel ignition occurred. The use of clamps has now decreased, owing to the difficulty in controlling these kilns and the high wastage involved.

Continuous kilns

These are based on the Hoffman kiln and comprise a closed circuit of about 14 chambers arranged in two parallel rows with curved ends (Fig. 4.3). Divisions between the chambers are made from strong paper sealed with clay and, by means of flues, the fire is directed to each chamber in turn. Drying is carried out prior to the main firing process and is achieved by warm air obtained from fired bricks during cooling. The kilns are described as continuous, since the fire is not extinguished – it is simply diverted from one kiln to the next, the cycle taking about one week. Coal was traditionally used but firing now may be by oil or gas. These kilns are very widely used for brick production.

Tunnel kilns

These are the most recently introduced kilns and can reduce firing time to little over one day. Units are specially stacked onto large trolleys incorporating a heat-resistant loading platform. The trolleys are then pushed end-to-end into a straight tunnel with a waist that fits the loading platform closely to prevent heat escape downwards. The bricks pass successively through drying, firing and cooling zones, firing normally being by oil or gas. The process provides a high degree of control over temperature, so that the process is suited to the production of high-strength, dimensionally accurate bricks. Perforated bricks are often fired in this way.

Properties of clay bricks

Strength

Since bricks are invariably used in compression, the standard method of test for strength involves crushing the bricks, the direction of loading being the same as that which is to be applied in practice – normally perpendicular to their largest face. Attention is drawn to the following aspects of the BS 3921 test (see Experiment 4.4).

1. The treatment of frogs or perforations should be designed to simulate their behaviour in practice – hence the perforations in perforated bricks should not be filled prior to testing and frogs should only be filled if the bricks are to be used 'frog up'. When frogs are filled, the strength of the mortar used should be in the range 28–42 N/mm^2 at the time of testing.
2. Strength varies with moisture content and, since the saturated state is the easiest state to reproduce, bricks should be saturated for 24 hours (or by vacuum or boiling) before testing.
3. In order to obtain representative results, 10 bricks should be tested, these being carefully sampled either during unloading or from a stack.
4. Bricks are tested between 3 mm thick plywood sheets, which reduce stress concentrations resulting from irregularities in the brick surfaces.
5. The strength of the brick is calculated as:

$$\frac{\text{maximum load}}{\text{area of smaller bed face}}$$

(The two bed faces may not always be identical in size.)

Bricks may be designated as Class 1, 2, 3, 4, 5, 7, 10 and 15, according to their average compressive strength. These classes derive from formerly used Imperial units, the numbers being multiplied by 6.9 to give the minimum crushing strength in N/mm^2. Hence, a class 10 brick has a minimum compressive strength of 69 N/mm^2.

There is normally good correlation between density and strength, though the precise relationship depends on the method of forming and firing the brick. Underfired bricks would have much reduced strength for a given density, as would bricks which are damaged by firing too rapidly.

Water absorption

Water absorption may be an important property of clay bricks, since bricks that have very low absorption are invariably of high durability (though the converse is not always true).

In the BS water absorption test (Experiment 4.5) oven-dried bricks are either boiled for 5 hours or subjected to a vacuum test in order to ensure the maximum possible penetration of water (even in these tests there will normally be some voids which remain unfilled). An alternative method, for control purposes only, is to soak the brick in water for 24 hours, though this gives a lower result than the first two tests.

The formula for water absorption, given as a percentage, is:

$$\frac{\text{mass of water absorbed}}{\text{mass of oven dried brick}} \times 100$$

Efflorescence

This is the name given to the build-up of white surface deposits on brickwork when drying after earlier saturation (Fig. 4.4). It results from dissolved salts in the brick or other sources and quite commonly spoils the appearance of new brickwork. It will be appreciated that brickwork often reaches high levels of saturation during and immediately after construction for one or more of the following reasons:

Fig. 4.4 *Efflorescence in new brickwork.*

- Bricks may not have been fully protected from rain prior to laying.
- Mortar loses moisture into the bricks by suction.
- Brickwork is readily saturated by weather prior to protection by a roof.

Drying of buildings takes place immediately after roofing and is accelerated when internal heating is commenced or in good weather conditions. In external walling drying occurs mainly in warm, dry weather; thus, in both cases, efflorescence tends to be worse during the spring period. Efflorescence can be difficult to remove; it does not respond well to brushing, and wetting the wall tends to redissolve the salts allowing them to be re-absorbed into the wall. In most cases efflorescence is a temporary problem which will disappear after a few seasons. It is unfortunate that this problem is often at its worst when the building is being handed over to the client!

An efflorescence test is described in the 1985 edition of BS 3921. In a 1995 amendment to the standard the test was deleted because: 'it has proved difficult to relate the result of the test to the liability of the brickwork to produce efflorescence'. (One of the problems is that salts may also be drawn from the sand used for the mortar, or from the ground.)

To guard against excessive soluble salts in *any* clay brick in the absence of the efflorescence test the soluble salt content requirement (see below) is now applicable to all clay bricks. Since the concept of efflorescence is still applicable to clay bricks the test is included as an experiment (Experiment 4.3), though it no longer has BS validity.

Soluble salt content
Salts can be harmful in

1. Causing efflorescence.
2. Leading to problems such as sulphate attack in the mortar.

Bricks are classified as of low (L) soluble salt content if the percentages of the following ions do not exceed the levels stated:

Sulphate	0.5%
Calcium	0.3%
Magnesium	0.03%
Potassium	0.03%
Sodium	0.03%

The figures for the calcium and total sulphate ions are higher because calcium sulphate has low solubility and is therefore less damaging than the other common sources listed. Bricks not achieving this standard must comply with the normal (N) soluble salt requirement, in which the percentages of the following ions do not exceed the levels stated:

Sum of contents of sodium,
potassium and magnesium: 0.5%
Sulphate 1.6%

British Standard requirements for engineering bricks

Engineering bricks have specific requirements relating to absorption and strength in addition to those such as dimensional tolerances and efflorescence requirements which apply to other types. Table 4.2 indicates these requirements for engineering class A and class B bricks. The water absorption values are based on the 5-hour boiling test.

Table 4.2 *Requirements for engineering bricks*

Class	Min compressive strength (N/mm^2)	Max water absorption (%)
Engineering A	70	4.5
Engineering B	50	7.0

Frost resistance

Frost resistance of clay bricks is designated as follows:

- **Frost resistant (F)**
  Bricks that are durable in all building situations including those where they are in a saturated condition and subjected to repeated freezing and thawing.
- **Moderately frost resistant (M)**
  Bricks that are durable except when in a saturated condition and subjected to repeated freezing and thawing.
- **Not frost resistant (O)**
  Bricks that are liable to be damaged by freezing and thawing if not protected as recommended in BS 5628: Part 3 during construction and afterwards – for example, by an impermeable cladding. Such units may be suitable for internal use.

Expansion on wetting of clay bricks

Clay bricks fresh from the kiln are desiccated due to the high temperatures experienced in the firing process.

On exposure to ordinary atmospheres fired clay products gradually take in moisture, their size reaching an equilibrium value in about six weeks depending on brick type and conditions.

This expansion is partly offset by mortar shrinkage but net expansions of around 300×10^{-6} (about 0.3 mm per m length) may still occur. Clay bricks should not therefore be used fresh from the kiln in order to allow this expansion to take place. If they are laid within this period there is a risk of expansion of the wall and consequent risk of buckling if that expansion is restrained by some external structure (such as a concrete frame). Perhaps surprisingly, immersing bricks prior to laying does not remedy the problem.

Durability of clay brickwork

The durability of clay brickwork is likely to be more of a problem than its strength, since in most situations clay bricks are very much stronger than is required structurally.

Almost all durability problems are associated with moisture penetration and it is therefore of paramount importance that bricks be suited to the degree of exposure likely to be found. In any one structure the most exposed position should be identified and, assuming that the same brick type is to be used throughout, the brick selected should be adequate for this degree of exposure.

The causes and mechanism of the three main modes of deterioration are considered below.

Frost damage

This (see Experiment 4.7) is associated with the crystallisation (freezing of water) below the surface of the bricks. Hence, for this to occur the bricks must be fairly porous in order to admit water. There is strong evidence to suggest that water contained in bricks moves into cracks or large voids or towards the surface as the temperature is lowered, damage occurring as ice crystals grow in any enclosed positions. Figure 4.5 shows ice crystals being extruded from pores and cracks in the brickwork. Damage is progressive, a few freezing cycles rarely causing visible damage, and it will normally take several years for frost damage to become apparent. The susceptibility of bricks to frost damage clearly increases as their water content rises, though some highly porous bricks are found to be frost resistant even when saturated.

The size distribution of pores in a brick rather than total porosity determines frost resistance.

Bricks containing coarse pore structure are generally more frost resistant than those with a fine pore structure. Figure 4.6 shows the fractured surfaces of a Fletton (type 'O')

Fig. 4.5 *Ice extrusions from pores/cracks in brickwork, subject to freezing when saturated.*

Fig. 4.6 *Scanning electron microscope photographs of fracture surfaces in (a) Flectton brick (type 'O') and (b) Otterham brick (type 'F').*

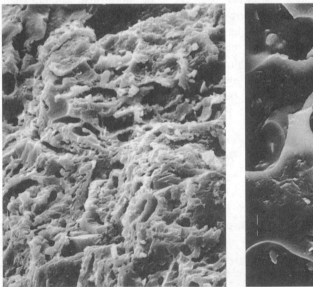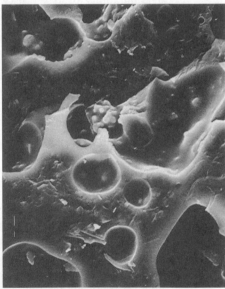

brick and an Otterham (type 'F') brick magnified about 300 times using a scanning electron microscope. Although the two brick types have similar absorption values of about 20 per cent, the degree of vitrification is seen to be larger in the case of the Otterham brick, resulting in fewer fine spaces within the brick and therefore a more frost-resistant product. However, since pore size distribution is difficult to measure, frost resistance is best assessed from long-term experience. Perhaps surprisingly, some stock bricks that have quite low strength and high water absorption are found to be frost resistant, presumably because they contain a coarse pore structure.

As a general rule, the resistance of brickwork to frost will be greatest when all possible means are taken to prevent moisture penetration. In the case of types 'M' and 'O' bricks the highest risk may be present during construction. It is important to keep bricks dry and to avoid high moisture contents in partially complete structures which do not have the protection of a roof, especially during cold winter spells.

Crystallisation damage

This (see Experiment 4.8) is associated with the crystallisation of salts beneath the brick surface. The extent of damage is directly related to the soluble salt content of bricks, hence the best method of avoiding damage is to use type 'L' bricks. Salts from other sources may also contaminate brickwork. Such sources include poor quality sand in the mortar, contaminated ground-water and wind-borne spray from the sea. Nitrates or chlorides are more common than sulphates, though they all have largely the same effect. Some salts may even be released from the cement, though the effect of these should be short-lived.

One problem of salt crystallisation is that it tends to occur locally at boundaries between damp areas of brickwork, such as parapet walls, and adjacent drier areas. Salts in solution are drawn to these drier areas and deposits therefore build up, causing local damage. Many salts are hygroscopic (they absorb water), so that they may tend to perpetuate dampness where they occur. Hence, in order to avoid these problems:

- Use well-fired bricks such as type 'L' in situations where dampness is likely.
- Use clean materials and avoid contact between brickwork and ground-water or soil.
- Design and detail brickwork in such a way as to minimise exposure to moisture.

Sulphate attack

This occurs in cement mortars – especially those in which the cement contains appreciable quantities of tricalcium aluminate, though the sulphates responsible often originate in the bricks. Sulphate attack, which only occurs in damp situations, results in expansion and eventual disruption of the brickwork. Figure 4.7 shows disruption of the mortar joints in a clay brick retaining wall. Avoidance of sulphate attack is achieved by the same precautions as given under the section on crystallisation damage, though the use of sulphate-resisting cement in the mortar will reduce the severity of attack where these precautions cannot be met fully, such as in earth-retaining walls.

Calcium silicate (sand lime) bricks (BS 187)

These bricks are made by blending together finely ground sand or flint and lime in the approximate ratio of 10 : 1. The semi-dry mixture is compacted into moulds and autoclaved, using high-pressure steam for several hours. A surface reaction occurs between the sand and lime, producing calcium silicate hydrates which 'glue' the sand particles into a solid mass. The main properties of calcium silicate bricks are:

- A high degree of regularity, with a choice of surface texture ranging from smooth to 'rustic'.
- A wide range of colours, produced using pigments.
- Very low soluble salt content, so that efflorescence is not a problem.

Fig. 4.7 *Sulphate attack in clay brickwork retaining wall. The sulphates originated both in the bricks and in the ground (the wall has no DPC).*

Table 4.3 *Calcium silicate bricks: strength classes and applications*

Calcium silicate brick strength class	Minimum mean compressive N/mm²	Use
2	–	Protected external applications
3	20.5	Free-standing external, or below dpc
4 or greater	27.5	Copings and retaining walls

- Fairly high moisture movement.
- Compressive strength in the range 7–50 N/mm². (Strength classes are employed in BS 187.)
- Good overall durability in clean atmospheres, though they may deteriorate slowly in polluted sulphur-containing atmospheres.

Selection should, as with clay bricks, be according to purpose but, unlike clay bricks, strength is the best guide to durability. The uses of calcium silicate bricks are shown in Table 4.3.

Calcium silicate bricks may exhibit substantial shrinkage. An indication of this is found in BS 187, the 1978 edition requiring that shrinkage does not exceed 400×10^{-6}, which is of the same order of magnitude as the initial *expansion* of clay bricks. (Note that this requirement of BS 187 was deleted by amendment in 1987 due to problems with the test.) On account of their increased moisture movement, joints in calcium silicate brickwork are recommended at roughly 7 m lengths, and they should not be laid wet. Since movement characteristics are different, they should not be bonded directly to clay brickwork. The use of calcium silicate bricks has increased considerably in recent years, due at least in part to the fact that they are now available in a variety of colours, including pastel shades of red and blue, as well as the original off-white.

Concrete bricks (BS 6073)

These have been introduced quite recently and comprise well-compacted low-workability concrete mixes of appropriate aggregate size, leading to products of high strength and durability. Colours and textures can be obtained which can give a final appearance very similar to that of clay bricks. They are free from efflorescence, tending to shrink slightly in dry situations. Owing to different movement characteristics, they should not be bonded to other brick types. With a limited number of production plants currently operating, transport costs may be high except in the regions where they are produced. There may also be problems of obtaining a good colour match if, at a later stage, extension or modification of buildings employing such bricks is undertaken. Specially shaped concrete bricks are now very widely used for paving, mainly of pedestrian areas. When compacted into a sand bed they provide an attractive, free draining surface with ready access for underground repairs if required.

4.3 Building blocks

These may be defined as building units having at least one dimension in excess of the maximum specified for bricks (see 'Bricks' page 116). They are designed to be bedded in mortar to produce walling or to be used as filler blocks in reinforced concrete floors. Blockwork has developed into a major form of construction for the following reasons:

- Rates of production during construction are substantially greater with blocks than with brick-size units – for example, a 100 mm thick block of size 440 × 215 mm is equivalent to approximately six standard bricks.
- A great variety of sizes and types is available to suit purposes ranging from structural use to lightweight partitions.
- Modern factory production methods ensure consistent and reliable performance.
- High-quality surface finishes are obtainable, obviating the need for rendering or plaster coatings in many cases.

Clay building blocks

Clay blocks are generally hollow units (formed by extrusion), since the voids result in:

- Reduced firing times
- Reduced weight
- Increased handleability
- Increased thermal insulation.

Fig. 4.8 *Clay building blocks.*

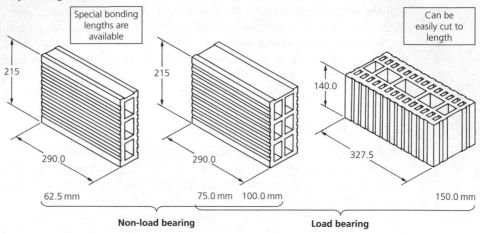

Figure 4.8 shows the various types of block available. Since blocks are made from well-sintered clay, the material is dense and quite brittle. The clay 'skin' is often quite thin so that cutting parallel to ribs is easily achieved by fracturing the clay with a trowel in line with them. Chasing may also be carried out by the same technique. Clay wall and partition blocks are keyed for rendering or plastering, as they are not intended to be used as finished facing materials. Masonry fixings must be applied with care since masonry drills may fracture thin clay skins.

Concrete blocks

Concrete blocks are covered by BS 6073, which also covers concrete bricks. They may be solid, hollow, or cellular.

- *Solid blocks* are largely voidless but may have grooves or holes to reduce weight or facilitate handling. These must not exceed 25 per cent of the gross volume of the block.
- *Hollow blocks* have voids passing right through. They can be 'shell-bedded' – that is, the mortar is laid in two strips adjacent to each face, so that there is no continuous capillary path for moisture through the bed. Loadbearing capacity is, of course, reduced when blocks are laid in this way. The strength of hollow blockwork can be increased by filling the cavities with concrete, especially if reinforcement is included. Sound insulation is also improved in this way.
- *Cellular blocks* are a special type of hollow block in which the cavities are closed at one end. The solid edge would normally be laid upwards and, in the case of thin blocks, this makes it easier to produce an effective bed joint.

A great variety of aggregate types is used in the production of concrete blocks – these include natural aggregates, air-cooled blast-furnace slag, furnace clinker, foamed, expanded or granulated blast-furnace slag, bottom ash from boilers and milled softwood chips. The density and strength of the resultant product vary accordingly and they are also influenced by the manufacturing technique.

Denser types of block can be used for loadbearing purposes, and decorative or textured facings may be applied.

Lightweight types are now in particular demand, owing to the need for high thermal insulation walling. Composite blocks (Fig. 3.25), which have a dense and durable decorative facing with lightweight backing for thermal insulation, are also available and many types include pulverised fuel ash and may be aerated and/or autoclaved. Higher insulation varieties of autoclaved aerated blocks are easily cut with hand saws and can be used without cavities to produce walling to current thermal insulation requirements. (The waterproofing function would be achieved by application of a suitable external rendering or cladding.)

BS 6073 does not lay down strength classes but gives minimum strengths for all block types:

- **Thickness not less than 75 mm**   Average strength of 10 blocks not less than 2.8 N/mm^2. No individual block to be less than 80 per cent of this value.
- **Thickness less than 75 mm**   Average transverse strength of 5 blocks not less than 0.65 N/mm^2.

(The strength requirement for thinner blocks is given as a transverse (bending) test because there would be a tendency for such blocks to buckle in a compression test.) BS 6073 also stipulates drying shrinkage, which should be:

- Not more than 0.06 per cent (except autoclaved aerated blocks).
- Not more than 0.09 per cent (autoclaved aerated blocks).

It is important, on account of their high shrinkage, that autoclaved aerated blocks are not saturated prior to laying, otherwise shrinkage cracking may result. Handleability is also much impaired when wet owing to their greatly increased weight in this condition.

No guidance regarding durability of blocks is given in BS 6073. The principles given in Chapter 3 on concrete apply, though it may be added that some open-textured or autoclaved blocks are often more resistant to frost than their strength would suggest. This may be due to the type of void present and the fact that saturation is rare. Nevertheless, where severe exposure or pollution is likely, blocks of average compressive strength not less than 7 N/mm^2 should be specified. All types can, of course, be protected by rendering.

The strength categories included in the standard (in N/mm^2) are :

2.8   3.5   5.0   7.0   10   15   20   35

It may be surprising that such low strengths are acceptable, but it should be appreciated that these correspond to lightweight aerated blocks with the best **thermal** performance. Loadbearing capacity of these blocks is adequate for low-rise construction; for example, a block of bedface dimensions 450 by 100 mm and strength 2.8 N/mm^2 would have a loadbearing capacity of:

2.8 × 450 × 100 = 126 000 N, which is 126 kN of 12.6 tonnes (approx.)

Although the strength of the corresponding masonry will be less than this, it is very adequate for most purposes.

4.4 Natural stone masonry

Use of natural stone masonry has diminished greatly during the twentieth century largely on account of increased labour costs which have made the processes of extraction and cutting/dressing of stone uneconomic. The subject is discussed briefly because there are many stone buildings still in existence, and some stone construction is still carried out in areas where stone is available from demolition of old buildings or where stone production continues, albeit on a limited scale.

The biggest barrier to continued major use is the very high labour costs in cutting or dressing the material. Stone is now mainly used as a cladding material since some types have excellent appearance and weathering properties and their use symbolises quality or prestige. Stone in this context is fixed as generally relatively thin slabs to a structural frame and therefore must be regarded as a cladding rather than the more traditional masonry form which will be summarised here.

The properties of stones vary greatly according to classification and also according to location. The performance of sandstones in particular can vary within wide limits according to the nature of the cementing material, which may be based upon calcium carbonate, iron oxide or silica. The best indication of likely performance is therefore obtained by reference to existing buildings from each specific materials type and origin.

Traditional stone masonry must be suitably bonded as any other form. The commonest method of construction is to construct corners ('quoins') from dressed stone and these are built first to define the building shape and plumb the façades. Thereafter stone is used to form infill panels and the quality of this may vary considerably. Figure 4.9 shows an example of roughly coursed stone which is relatively cheap to build and Fig. 4.10 shows closely cut stone used in conjunction with much finer joints. In earlier examples of stone construction the 'mortar' may have comprised mud or clay with a lime finish to internal and external surfaces for weather protection. Figure 4.11 (200-year-old dwelling, now decayed) shows that a satisfactory result could be obtained by

Fig. 4.10 *Close-jointed stone providing good interlock.*

Fig. 4.11 *Random uncoursed rubble walling 600 mm thick, held together by interlock and soil/rubble core. The wall is now decaying because the roof failed.*

random uncoursed rubble with soil infill. Larger internal voids would be filled with rubble. Such structures depended heavily on a roof to protect them from the effects of moisture. More recently the inner skin would be built from blockwork to form a rigid background for the stone and tied at each storey height by a reinforced concrete ring beam. Cavities are not normally found as they are difficult to execute unless fully squared stone is used. In modern stone construction, building regulations can be satisfied by incorporating insulation either within the wall or adjacent to the internal face. In Fig. 4.9 the 200 mm stone facing is butted against 200 mm of extruded clay blockwork with reinforced concrete ring beams at each storey. Insulation is provided in the form of 70 mm of fibreglass bonded to dry lined internal plasterboard. In Fig. 4.10 the wall comprises 200 mm of stone followed by 70 mm of rigid polystyrene and 105 mm of extruded clay block. The stone is tied through to the blockwork.

The walls in each of the figures resist moisture penetration by the 'overcoat' effect referred to in BS 5628: Part 3. The stones are absorbent, thus wind-driven rain is absorbed by the material rather than penetrating through the wall. This moisture is then progressively released by evaporation during fine weather.

Absorbent stones, such as those of sedimentary origin, should not be bonded or used in close proximity as they will probably have different movement properties. Limestone, in particular, should not be used above softer sandstones, as calcareous matter could be washed out and later crystallise in the sandstone, causing disruption.

The structural performance of stone masonry depends greatly upon the bond used. BS 5628 allows bearing strengths for close-bonded masonry (as in Fig. 4.10) to be derived from the strength of the stone itself rather than from the mortar, allowing, of course, for shear/slenderness considerations. Where random rubble masonry is used, the overall strength may be as little as half that obtained using the standard design method. As a general rule, the mortar used for natural stone walling should be softer than the stone itself. The mortar therefore saturates first in wet conditions and, if freezing or crystallisation occurs, damage should be concentrated in the mortar. This can then be removed and the wall repointed, which is preferable to damage to the stone itself.

Table 4.4 gives a summary of origin, properties and applications of some stones still in common use.

4.5 Masonry mortars

Mortars may be defined as mixtures of sand, a binder such as lime or cement, and water. Their prime function in masonry is to take up tolerances between building units such as bricks or blocks, though they also have to satisfy some or all of the following requirements:

1. They should impart sufficient strength to the complete unit.
2. They should permit movement (unless this is negligible or joints are provided). When movement occurs within a well-constructed masonry unit, it should take place in the form of microcracks within the mortar rather than cracking of the bricks or blocks.
3. They should be durable – that is, they should resist frost or other forms of environmental attack.
4. They should resist penetration of water through the unit.
5. They should contribute to the aesthetic appearance of the wall.

To permit effective use, mortars should be workable, yet cohesive in the fresh state; bricks or blocks should need only minimal effort to bed them in the correct position.

Functions 1, 3 and 4 above are best provided by a strong mortar, though the wall as a whole will be in the region of five times stronger than the mortar on account of the

Table 4.4 *Common natural stone types: properties and applications*

Type	Sub-type	Properties	Applications
Igneous (once molten)	Granite	These stones are virtually pore-free and composed of mainly silica crystals with impurities resulting in a coarse, sometimes brightly coloured grain structure. They are immensely strong and durable but difficult to extract and cut as they have no natural bedding planes. They can be polished to a lasting shine	Formerly widely used in northern cities such as Aberdeen. Still used in reconstituted stone and in thin claddings to prestige city buildings
Sedimentary		All sedimentary rocks have bedding planes which facilitate extraction but limit the thickness of stone obtainable and may cause weakness through water admission. (Lay bedding planes horizontally where possible)	
(Cemented sediments of glacial origin)	Soft sandstone (abs. > 4%)	Lower extraction/cutting costs and easy to work to a fine finish; but may suffer from frost damage or pollution, especially if water penetrates bedding planes	Widely used in Cotswold towns/ Oxford colleges and parts of Scotland
(Cemented sediments of glacial origin)	Hard sandstone (abs. < 4%)	These usually have a coarse grain structure and therefore coarse surface texture. They may have high strength. Some varieties are extremely durable – for example, York stone	Formerly widely used, mainly in northern England towns for civic and general building
(Cemented sediments of marine life origin)	Limestone	As for soft sandstone. In addition, since limestones comprise mainly calcium carbonate, they are vulnerable to attack from pollution – particularly sulphur-based pollutants which convert the surface to calcium sulphate which flakes off due to differential movement	The best known are Portland stones (Dorset). Widely used in London and southern England
Metamorphic (normally sedimentary modified by heat/pressure)	Slate	Properties variable according to origin but the best varieties are very durable. They have many closely spaced bedding planes which permit cutting into fine sheets	Formerly widely used for roofing and DPCs. Many buildings in the lake district are constructed from solid 'green' slate

Table 4.5 *Common mortar types: properties and constituents*

Data taken from BS 5628: Part 3

		Mortar designation	Type of mortar (proportions by volume)		
				Air-entrained mixes	
			Cement/lime/ sand	Masonry/ cement/sand	Cement/sand with plasticiser
Increasing strength and improving durability	Increasing ability to accommodate movements due to temperature and moisture	(i)	1:0 to $^1/_4$:3		
		(ii)	1:$^1/_2$:4 to 4$^1/_2$	1:2$^1/_2$ to 3$^1/_2$	1:3 to 4
		(iii)	1:1:5 to 6	1:4 to 5	1:5 to 6
		(iv)	1:2:8 to 9	1:5$^1/_2$ to 6$^1/_2$	1:7 to 8
		(v)	1:3:10 to 12	1:6$^1/_2$ to 7	1:8

Direction of change in properties is shown by arrows

→ Increasing resistance to frost attack during construction

← Improvement in adhesion and consequent resistance to rain

very small mortar thicknesses used – typically 10 mm. Function 2 is best satisfied with weak mortars which crack readily if movement occurs. This leads to the working rule:

The mortar must not be stronger than the units it is bonding.

Table 4.5 shows five types of mortar in common use together with compressive strength, durability and movement characteristics. Types (i) to (iv) are classified in BS 5628: Part 3.

For high-strength units such as engineering bricks, type (i) can be used, the cement content being almost sufficient to produce a cohesive mortar without additional plasticisers. These mortars will give great strength and durability and can be used in situations of high exposure and high stress, though separate movement provision may be required. Some 'softer' sands may not give adequate durability with type (i) mortars, since they are too fine and lead to high water requirements. Coarser sands to BS 1200 or finer concreting sands to BS 882 should be suitable.

Types (ii) and (iii) are suitable for general masonry work in winter. They have lower strength than type (i) but the lime acts as a plasticiser in the fresh material and helps bond the units and resist water penetration in the hardened material. Types (iii), (iv) and (v) mortars contain progressively more lime to compensate for the reduced cement content in plasticising terms. Type (iv) mortars can be used for general purposes in summer, while type (v) mortars are restricted to internal use because they have low durability. Note that when bonding weak products, such as lightweight blocks externally, low-strength mortars are preferred – protection would normally be provided by a cladding or rendering. Where low-strength stock facing bricks are used, a higher strength pointing can be used for protection, though such pointing is not recommended in highly stressed situations; it tends to transmit stress through the brick faces, being stiffer than the remaining material, and this may lead to spalling of the brick surface.

Use of plasticisers

Plasticisers fulfil a most important function in fresh mortar. The material needs to be highly workable – in fact an 'ideal' workability might be considered to be that level of fluidity at which the masonry units need minimal force to bed them correctly, and their own weight should be almost sufficient for the purpose. But they must also be **cohesive**, since wet mortars would be prone to moisture loss or 'bleeding'. The plasticiser enables these two functions to be fulfilled at the same time.

Lime is in fact hydrated lime, $Ca(OH)_2$, normally sold as a bagged powder which eventually contributes slightly to strength by carbonation – see the equation on p. 89. Most limes are non-hydraulic – they do not set by addition of water – though some limes containing impurities such as magnesium carbonate, as obtained in dolomitic limestone, do have some hydraulic action. Lime plasticises mortar very effectively, leading to a workable yet 'fatty' material. Lime mortars are now less widely used than previously since lime is bulky and unpleasant to handle, though in some situations, for example, in exposed conditions with low-suction bricks, their use is recommended to help to avoid water penetration.

Instead of lime, liquid plasticisers may be used (see Table 4.5), having an air-entraining effect and, therefore, leading to improved frost resistance, though mortars so produced are less cohesive in the fresh state and less impermeable in the hardened state. Recent research has shown that excellent 'all-round' performance is obtained by incorporating a liquid plasticiser in a $1:1:5\frac{1}{2}$ cement/lime/sand mortar – leading to the advantages of both lime and liquid plasticisers.

A further alternative is supplied by *masonry* cements. In these types plasticisers are incorporated during manufacture. They have properties intermediate between the above types (see Table 4.5).

Aesthetic considerations

As well as the functional requirements described above, mortars should contribute to the aesthetic quality of the walling. The commonest problem is associated with colour; to achieve the best appearance, uniformity of colour from batch to batch is essential. This requires consistency of colour of sand, cement and, if used, pigment, together with accurate batching. Sand and cement should be obtained in adequate quantity to ensure uniformity during construction. Sand quality is often variable and, although unwashed 'soft' (building) sands are generally suitable, they should be checked for excessive clay content or fine material (BS 1200). To achieve repeatable mixes, the use of gauging boxes for batching is recommended. Machines are now available for automatic batching and mixing of mortars and these should give successive batches of very uniform quality and colour. A further option is the use of ready-to-use mortars in which case the supplier should guarantee consistency of colour. These mortars are set retarded so that they can be used up over a working day.

Thermal considerations

There is concern that, with steadily increasing demands for good thermal insulation, the mortar joints cause quite extensive cold bridging, especially in solid wall construction, increasing heat transmission by as much as 25 per cent. This can be overcome by the use of recently introduced lightweight mortars in which sand is replaced by a suitable crushed lightweight aggregate, such as perlite. It should be possible by careful mix design, to satisfy the above functional requirements with the much lower heat transmission which is characteristic of lightweight materials.

Shrinkage

As a final cautionary note, it should be appreciated that mortars exhibit very high shrinkage, often over 2000×10^{-6}. This is of little consequence when correctly used in small thicknesses (up to 15 mm) since a network of fine cracks results. Mortars should not, however, be used to fill large spaces, especially in one operation; concrete or other materials with relatively low shrinkage would perform such functions much more satisfactorily.

4.6 Strength of masonry

Comments were made in Chapter 1 about the need to interpret compression test results with caution, and it would certainly be unwise to assume that the compressive strengths of masonry would be of similar levels to those of the units used to build it. Even ignoring slenderness considerations, the compressive strength of masonry is usually substantially less that of the units it comprises. This is due to the shape of the units and is an extension of the argument given on page 9. The risk of shear/tensile failure in the vertical plane is much higher than that in bricks or blocks when tested individually. For this reason the mortar strength does not have a large effect on masonry strength, especially at the lower end of the strength range, since the mortar bed is very thin. The situation is illustrated by figures in Table 4.6 for masonry strength as a function of mortar and brick strength.

Table 4.6 *Masonry compressive strength*

Relationship between brick and mortar strength

Characteristic strength of brick (N/mm²)	Mortar type/strength	
	Type (iv) (1.5 N/mm²)	Type (i) (16.0 N/mm²)
20	5	7
70	11	19

It will be noticed that with the weaker brick the strength of the masonry is much higher than that of the weaker mortar and is not greatly affected by the mortar strength. This situation prevails in most common construction and the mortar strength is dictated by durability rather than by strength requirements.

In the case of the stronger (engineering) brick, the strength of the mortar has a much more pronounced effect on masonry strength.

4.7 Renderings

These may be defined as cement-based mortar coatings for external or other surfaces which must be water resistant. They have the following chief functions:

- To provide an aesthetically pleasing, easily maintained surface finish.
- To resist rain or damp penetration.

To achieve these functions, rendering must

- Adhere well to the background.
- Be reasonably impermeable and free from cracks.
- Have adequate durability.

The mortars used for rendering are basically the same as those for brickwork, though careful attention should be paid to sands (see BS 1199). Coarse sands will lead to mortars with poor adhesion while very fine sands or sands containing clay require more water and lead to shrinkage problems. As with jointing mortars, the mortar should not be stronger than the background. Lime is again considered a valuable material, especially in weaker mortars.

The correct technique for rendering involves the application of at least two coats.

The principles of 'render, float and set' apply as in plastering (see p. 175) though in most cases two-coat work is sufficient. The first coat is designed to level the background and to even out its suction or water absorption, which may vary considerably from place to place, being lower with lintels, mortar joints and dense bricks and higher with lightweight blocks. Treatment of the latter with PVA (polyvinyl acetate) emulsion may be necessary to reduce suction to acceptable levels. Saturation with water is not recommended as it leads to later shrinkage. Raking of mortar joints is essential for good adhesion if the render is to be applied to a smooth background such as common bricks. The 'ideal' thickness for the undercoat is about 15 mm. Adequate trowelling in the undercoat is very important; this compacts the mortar, closing cracks as the water is drawn into the background. Trowelling would normally be completed within about one hour of application. The undercoat is then 'scratched' to assist in producing a uniform distribution of small cracks and to provide a key for the final coat. Curing for at least three days is necessary; a weak and friable product results if there is inadequate water for cement hydration. Premature drying is easy to spot since the material becomes much lighter in colour as it dries. After curing, **drying should be allowed** in order to permit a network of fine shrinkage cracks to form.

Fig. 4.12 *Stucco finish to buildings in Brighton. Lower elevations are ruled to produce an ashlar (coursed stone) appearance. Scrolls and other decorative details are moulded.*

The final coat should be applied after the first coat has cured and dried. The ideal time between coats is about one week. The final coat should always be thinner than the undercoat. It covers cracks in the former, and any cracking should be very fine since, in low thicknesses, shrinkage is restrained by the now fully shrunk undercoat. The final coat may take several forms:

- **Smooth rendering**
  A smooth finish is obtained by trowelling, with a wooden float, to about 3 mm thickness. Metal trowelling produces a very smooth finish but is more likely to be affected by crazing problems. Smooth rendering can be shaped and ruled to produce the appearance of natural stone at a fraction of the cost; this is known as 'stucco' (Fig. 4.12).

- **Pebble dashing**
  Pebbles of size 3–10 mm are dashed into the final coating while plastic. This provides a highly textured surface which tends to hide blemishes or fine cracks. The final colour can be varied by selection of appropriate pebble type.

- **Rough cast**
  Mortar is dashed against the wall producing a textured finish. The coarseness of texture increases with the maximum aggregate size.

Some cracking invariably occurs in the final coat but is obscured in coarse-textured finishes. In smooth-rendered finishes, a very fine crack network results which should be invisible if the final coating is only a few mm thick. Smooth renderings are normally finally decorated with a masonry paint which will fill and obscure such fine cracks. Curing of the final coat is essential in all cases for best results.

The production of good-quality renderings requires skill and experience as well as organisation of work; for example, the final coat should be applied in one continuous operation to each elevation in order to avoid unsightly joins in the material.

4.8 Dampness in masonry

A great many of the problems which commonly occur in buildings are associated with the presence of moisture, and some effects on specific materials have been outlined above. A large range of difficulties could arise and these include fungal attack, swelling/distortion, loss of strength, frost damage, sulphate attack, crystallisation, efflorescence, corrosion of metals and damage to finishes. It is for this reason that great care needs to be given to design, detailing and construction. Those buildings which shed rain-water (including wind-blown rain) effectively and prevent admission of ground moisture will be largely resistant to the problems listed.

The extent to which water is admitted and retained by the building fabric depends not only on the amount of water that is incident from the exterior of the building (although obviously good design will minimise this), but also on the surface characteristics of the materials used. As the most common means of admission of water is by capillarity in porous materials, both these terms will therefore be considered.

Porosity

Porosity relates to the total voids in a material. It can be defined as

$$\frac{\text{volume of pores in a given sample}}{\text{bulk volume of sample}} \times 100$$

or

$$\frac{\text{bulk volume of sample} - \text{solid volume of sample}}{\text{bulk volume of sample}} \times 100$$

Bulk density (D_B) and solid density (D_S) can be related as follows:

$$D_B = \frac{M}{V_B} \qquad \text{where } M \text{ is the mass and } V_B \text{ is the bulk volume}$$

$$D_S = \frac{M}{V_S} \qquad \text{where } V_B \text{ is the solid volume}$$

Therefore, porosity (P) is:

$$P = \frac{V_B - V_S}{V_B} \times 100 = \left(1 - \frac{V_S}{V_B}\right) \times 100$$

Hence

$$\frac{V_S}{V_B} = \frac{1 - P}{100}$$

Substituting:

$$D_B = \frac{M}{V_B} = \frac{D_S V_S}{V_B} = D_S\left(\frac{1 - P}{100}\right)$$

In the case of materials of fairly low porosity (say 10 per cent or less) the equation indicates that the percentage difference between the bulk density and the solid density of a material is equal to its percentage porosity. For example, a brick with bulk density 1500 kg/m³ and porosity of 8 per cent would have a solid density of 1.08 times 1500 = 1620 kg/m³.

Capillarity

In order to explain this term, it will be helpful to consider first the term 'surface tension'. Surface tension is exhibited by almost all materials as a result of cohesive forces which exist within them. These are intermolecular forces which arise naturally, owing to the electronic nature of the outer shells of atoms. In a liquid, the forces are relatively small and hence they only exert a slight restriction on mobility. Liquids falling freely through space nevertheless take up a spherical shape, since molecules at the surface of the liquid experience a net inwards attraction (Fig. 4.13). The surface is said to be in 'tension' because the inward attraction results in the body of liquid trying to minimise its surface area, hence producing a sphere – the same shape that would be produced if the surface of the liquid were covered by a thin elastic membrane.

Fig. 4.13 *Molecular cohesion – the origin of surface tension.*

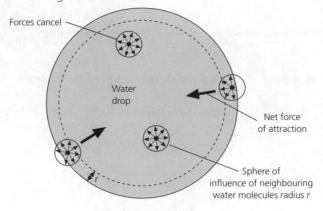

Forces cancel

Water drop

Net force of attraction

Sphere of influence of neighbouring water molecules radius *r*

Fig. 4.14 *Non-wetting properties resulting when cohesive forces exceed adhesive forces.*

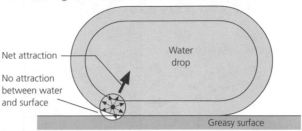

Solids also exhibit surface tension – in fact, the cohesive strength of solids can be likened to a sort of surface tension but without the flexibility which exists in liquids. In gases, surface tension effects are very small.

At an interface between a liquid and solid, liquid molecules will be under two influences:

- Cohesion within the liquid tending to lead to surface tension.
- Adhesion between the liquid molecules and the solid, caused by the same type of electronic attraction which results in surface tension.

The behaviour of the liquid will depend on the relative magnitude of these two effects. If cohesion within the liquid is greater, it will still tend to minimise its surface area. This occurs when, for example, mercury contacts most solids or when water rests on a greasy surface (Fig. 4.14). When adhesion between the liquid and solid is greater than cohesion within the liquid, the liquid spreads out on ('wets') the surface, forming only a very thin layer. Reducing the surface tension (cohesion) of water – for example, by adding detergents – increases the relative magnitude of the attraction. Water wets most building materials, including ceramics such as brick, stone and concrete, wood, metals and some plastics. This attraction is described as capillarity and can occur theoretically in all these materials.

An important factor affecting capillarity in practice is the size of pores associated with individual materials. The effect may be explained by consideration of the water rising up capillary tubes (hence the term 'capillarity') of various diameters (Fig. 4.15(a); Experiment 4.1). It is well known that the capillary rise in glass tubes increases as the tube diameter decreases (Fig. 4.15(b)), because the force exerted on the liquid at the interface increases in relation to its cross-sectional area as the tube diameter decreases. It is true that the total capillarity force involved reduces as the tube gets narrower, since the length of meniscus decreases but the area of the liquid supported decreases at a greater rate, so the result is an apparently higher attraction.

In summarising, it may be stated that if water wets a solid surface it will be drawn into any gaps or pores in that material, the attractive force increasing as the width of gaps or pores decreases. Water can in fact be drawn with some force into cracks or pores which are too small to be seen with the naked eye. For the same reason, such water will also be slow to evaporate from such materials once the source of dampness is removed, since it is held in position by the same capillary forces that caused admission initially.

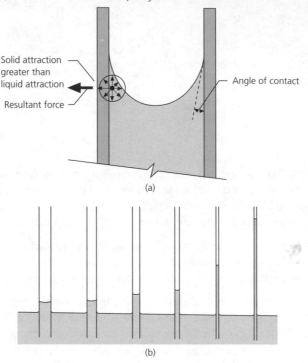

(a)

(b)

Interconnection of voids and pore structure

Capillarity can clearly only take place in materials which have interconnected voids. Some voids in most materials are not easily accessible to water and therefore cannot be easily saturated. Taking timber by way of example, the solid density of timber is quite high (approximately 1500 kg/m³), so that, if voids filled readily with water, timber would sink quite quickly on immersion. In fact, even low-density wood such as balsa wood, with its high void content, can float for some time, indicating that not all voids are readily saturated. Some aerated concrete blocks will also float on water for a time, again indicating that voids are not easily filled. The voids in air-entrained concrete are formed within the cement mortar and are therefore highly resistant to water penetration.

The importance of pore size distribution has also been mentioned in Section 4.2. It appears that although coarse pores greater than, for example, 30 nm diameter (1 nm = 10^{-9} m) may be readily saturated, they do not constitute a frost hazard because, on freezing, water tends to flow out of them to surface regions. The main problem occurs with smaller pores from which water tends to flow at reducing temperature, to 'feed' growing ice crystals.

At the present time, no simple method has been found of relating weathering resistance to pore structure, and, consequently, materials specifications for durability must be based upon other aspects of materials properties which relate to durability rather than to such terms as porosity. It was stated earlier that, in the case of clay bricks for example, some types of brick which are highly absorbent nevertheless satisfy the frost-resistance

Table 4.7 *Moisture penetration of masonry: possible problems*

Effect of dampness	Materials involved	Result
Fungal attack	Embedded timber	Softening and eventual failure
Swelling/distortion	Brickwork	May cause buckling or cracking if inadequate movement joints
Frost damage	Brickwork, concrete, mortar	Spalling followed by crumbling and eventual failure
Sulphate attack	Concrete/mortars	Disruption followed by failure
Crystallisation	Clay brickwork	Disruption followed by failure
Efflorescence	Brickwork, plaster and concrete	Unsightly appearance
Corrosion of embedded metals	Mainly embedded steel (aluminium if in contact with cement)	Loss of metal, followed by failure. (Examples, wall ties, steel reinforcement in poor concrete, screws, nails, etc.)
Damage to finishes	Wallpaper, emulsion paints, fabrics	Discolouration, mould growth, loss of adhesion
	Oil paints on cement substrate	Saponification (alkali attack)

requirement of BS 3921. In some cases, an absorbent brick can indeed increase the durability of the wall it is part of – for example, the resultant 'suction' of the brick can increase the cement mortar bond and thereby reduce the likelihood of capillarity at the brick/mortar interface. An absorbent brick will also tend to reduce the moisture content of the mortar joint – often the most vulnerable part of the wall (see Experiment 4.7).

A related property of materials is **permeability**, which is a measure of their ability to transmit water vapour or other fluid materials under pressure. The term is very relevant in sheet materials such as vapour barriers or coatings such as paints. The term *vapour resistance*, defined as

$$\text{vapour resistance} = \frac{\text{thickness}}{\text{permeability}}$$

is also used where referring to films of specified thickness. As an example, the vapour resistances of various films/coatings to water vapour are:

Film	*Vapour resistance* (GN s/kg)
Microporous paints	1–4
Gloss paints	5–8
0.15 mm polythene film	350
Aluminium foil	4000

Water admission can be reduced by such films in appropriate situations. Table 4.7 summarises some of the problems that may arise in masonry into which moisture has penetrated.

Prevention of moisture penetration in masonry

The following are intended to illustrate the various mechanisms by which moisture may penetrate masonry.

Cavity walls

There should be no path for moisture between the outer and inner skins. Hence cavities must be kept free of mortar droppings, vertical DPCs should be provided in reveals and wall ties should be designed to shed water. In exposed conditions there may be some risk of damp penetration even when a cavity is correctly installed. Such situations involve a combination of factors such as

- Tall buildings.
- High ground or absence of protection by other buildings or trees.
- Regions of high rainfall/average wind speed.
- Elevations facing the prevailing wind.
- Marine environments.

BS 5618 gives information on the assessment of exposure level in such cases.

When buildings are highly exposed, the design of masonry should bear in mind the factors affecting rain penetration given in Table 4.8.

Table 4.8 *Factors affecting risk of rain penetration in masonry*

Factor	High resistance to penetration	Low resistance to penetration
Brick or stone type	High absorbtion: abs. coeff. say over 20% (overcoat effect)	Low absorption
Mortar type	Includes lime	Other types
Mortar mix	Rich, e.g. 1:3 sand/cement	Weak, e.g. 1:9 sand/cement
Pointing	Weathered	Recessed
Cavity size	75 mm	25 mm
Cladding	Tile hanging	None

Cavity fills

Preformed types, which leave at least 50 mm of the cavity unfilled and are used with purpose-designed wall ties, are to be preferred.

Complete bridging of cavities with plastic foams or other materials for thermal insulation purposes will increase the risk of moisture penetration in exposed areas (see BS 5618). Insulants are classified as:

Fig. 4.16 *Weatherproofing details in parapet wall (additional insulation may be required).*

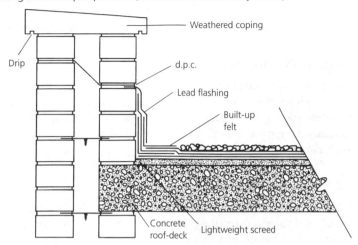

- **Type A**, including mineral fibre and polystyrene beads which should give performance comparable to an unfilled cavity.
- **Type B**, such as urea–formaldehyde (UF) foam and granular plastic fills, which carry greater risk of moisture penetration especially in situations as described in Table 4.8.

While type B insulants often provide greatly improved insulation at competitive cost, quality of workmanship is of paramount importance and some form of accreditation such as an Agrement certificate is recommended if potential problems are to be avoided.

Building design

In addition to the points raised above, the overall building design can have a major influence on moisture penetration. To reduce moisture penetration into masonry:

- Provide adequate eaves projection to shed water clear of the brickwork at roof level.
- Make provision for shedding of water from glazing, especially if used in large areas. Generous sill projections assist in this.
- Use damp-proof courses correctly positioned at low level and at high level for parapet walls (Fig. 4.16).
- Used weathered (sloping) coping where tops of walls are exposed.

Diagnostic measures when damp has affected masonry walls

Before measures can be taken to rectify the effects of dampness, it is essential to identify the cause of the problem. This should be done as soon as possible after dampness is detected since salts are often present in masonry, soil or adjacent materials, which tend to migrate to points of drying around any sources of dampness. If investigation is not carried out promptly, salts can build up to a level whereby they will seriously impair drying even after the fault has been rectified. The first signs of dampness are, quite often, deterioration of internal finishes, usually associated with damage to plaster resulting from migration of salts. The problem is often ascribed to rising damp, but before considering this source, other sources should be eliminated since treatment of rising damp may involve considerable expense and inconvenience. Possible alternative sources are detailed below.

Leaking rain-water system

Since very large quantities of water may be discharged from a roof. The importance of maintaining gutter and downpipe systems in good repair cannot be overstated. A simple problem such as a blockage due to leaves can result in large quantities of water being discharged down a wall face; only a wall of very high quality would be unaffected in such situations.

Leaking chimneys

Chimneys are highly exposed structures so that effective damp-proofing measures must be taken during construction. These include the use of strong sulphate-resistant mortars to resist penetration of water at mortar joints, together with suitable flashings/soakers. Poorly applied renderings can trap water rather than resist its ingress. Penetration of water down the chimney can cause problems, especially at upper-floor level.

Condensation

Since water vapour levels are generally higher inside dwellings, water has a tendency to diffuse outwards through the walls. 'Cold' internal wall surfaces are very easily created by, for example, positioning wardrobes on outside walls, water vapour then condensing on the wall, behind the wardrobe. This situation will be exacerbated if vapour levels in the property are already high. Large numbers of house plants are one common cause of high vapour levels. The risk of condensation problems can be estimated quite accurately. Relative humidity recorders are used to quantify the moisture levels present. The temperature of air and affected wall surfaces can also be recorded – preferably over several weeks. By use of psychrometric charts, the dew point of the air can be calculated, enabling predictions of condensation risk to be made. Significant dampness will result if the wall surface or underlying layer temperatures are below the dew point of the air for appreciable periods.

DPC or cavity bridging

DPCs can quite commonly be bridged by, for example, cements, renderings or soil, while cavities can be bridged by debris in the cavity such as mortar on wall ties, or at the base of the wall. A further cause in older properties may be where alterations have been made leading to debris in the cavity. It should be relatively easy to check for such problems, visibly from the exterior or by using a borescope in a small hole drilled from the outer surface.

Defective damp-proofing properties of the wall

Many older properties do not have cavities, relying on the 'overcoat' effect, whereby water is absorbed by a porous wall material in wet weather, the water drying out in dry weather spells. Where damp is present in solid walls, especially forming more exposed parts of the structure, the possibility that the damp is caused by penetration from outside should be considered. The problem will clearly be worse in wet weather and will depend on wind direction. It may be exacerbated when dense non-absorbent bricks or stone are laid in a bed of soft mortar. A further factor may be moisture from other sources, such as walls having poor thermal insulation, leading to cold internal surfaces and surface condensation.

Rising damp

Where investigation has eliminated the causes above and where dampness is mainly at lower levels in the building, the possibility of rising damp may be considered. The term is used for moisture penetrating upwards through masonry rather than by the other causes. It is common in older properties which may not have an effective damp-proof course, or in newer properties in which the DPC has deteriorated. It is usually associated with softer bricks or mortar which are more prone to absorbing or transmitting moisture. These often have a finer pore structure which enables moisture to climb higher against gravity. Rising damp will tend to reduce in severity in summer when the water table is lower and the water has to overcome a larger gravitational pull. Dry soils in summer will also tend to exert a counter capillary attraction on the moisture.

Diagnostic tests
A variety of diagnostics test exists and although some of these can be expensive, their use gives a better prospect of identifying the problem and adopting the most appropriate remedy.

Moisture content
An estimate can be easily obtained by using a 'Speedy' moisture meter. A small amount of plaster is removed and a sample drilled out with a 9 or 10 mm masonry drill, at low speed to avoid heating the sample. The mortar should be sampled as it is likely to be wetter than brick or stone. Reject the first 10 mm of the drillings.

Alternatively, the sample may be weighed and oven dried at 100 °C.

Electrical moisture meters can be used for the same purpose, though readings will be exaggerated if, as is common, salts are present. Dampness levels corresponding to moisture contents of material less than 5 per cent should not cause serious problems.

Effect of salts
These tend to delay drying and may indeed be responsible for most of the dampness by attracting moisture from the air. They are termed 'hygroscopic' salts due to their affinity for moisture. The Building Research Establishment has developed a test to measure the effect of any salts present. Samples of the masonry are obtained by drilling at several heights as above, a few grams for each sample sufficing. They are weighed and placed in sealed containers over a saturated solution of common salt (sodium chloride) which generates a relative humidity of about 75 per cent. After a period of about 2 days the samples are weighed again and oven dried at 100 °C. The measured moisture content

based upon extracted weight and their 'hygroscopic' moisture content after 2 days conditioning are compared.

> *If the measured moisture content is higher than the hygroscopic value, there must be some additional source of moisture other than the salts.*

It would not be possible for the salts alone to lead to this level of moisture. Figure 4.17 shows a case in which the measured moisture content is higher than the hygroscopic moisture content at lower levels in the wall. This suggests a rising damp problem.

Chemical composition

The pH of the moisture in the wall could be obtained (by mixing a sample with water and using an indicator) and compared with that of the ground-water and rain-water. This may assist in identifying the origin of the moisture. Also possible, though more expensive,

Fig. 4.17 *Example of wall in which the measured moisture content is higher than the hygroscopic value at lower levels, indicating rising damp.*

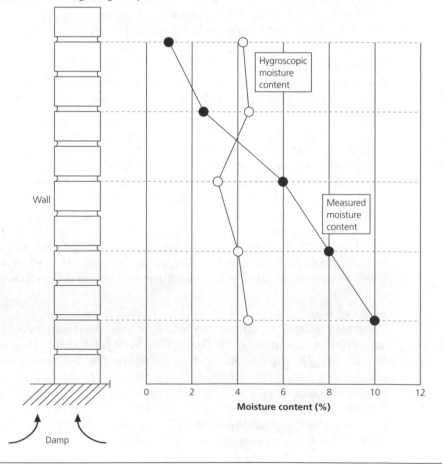

is to identify the chemical radicals of the salts and to compare these with those of the mortar in drier parts of the wall and with ground-water.

Remedial measures

For many of the above problems the following will be self-evident, though some of the areas merit further comment.

Rising damp, even if correctly diagnosed, can be quite difficult and expensive to rectify. A number of systems for rectification are available and these are given below in preferred order based upon research experience.

Most effective elimination of damp will be achieved by insertion of a mechanical DPC. Thin horizontal slots are cut using a chainsaw with special tungsten carbide teeth, at a level appropriate to the internal floor type. It is most important to lap new DPC sections and to ensure that each section is packed with mortar so that it supports the walls above. The procedure can clearly be quite expensive.

Alternatively, injection of silicones or aluminium stearate can be used to create a barrier to moisture. They act as repellents rather than as a positive moisture barrier, though once 'fixed' by chemical reaction with the mortar they should prevent moisture flow due to all but positive water pressure. Their effectiveness may be reduced by alkalis in newer cement-based mortars.

Other options include the use of ceramic tubes which draw water towards them, subsequently allowing it to evaporate; and 'electrical' methods.

It is advisable in every case to ensure that techniques have appropriate certification, and that they are carried out by a competent body, with comprehensive long-term guarantees.

Localised repairs to wall cavities containing obstructions are possible though the appearance of the masonry unit may be affected by such repairs. The repair of larger areas is likely to be more difficult. Use of surface-applied water-repellent coatings has been tried but it is difficult to prevent water penetration in brickwork by such techniques because most brickwork contains small cracks which readily admit water by capillarity even after treatment. A possible solution to rain penetration would be rendering (to which, if there is a history of problems, a waterproofing admixture should be added). Where renderings are defective they should be replaced.

Where internal plaster has been affected it should be removed; this will also remove any salts which often accumulate in the plaster. The background can then be primed with a moisture-resisting form of PVA (polyvinyl acetate) emulsion to prevent any residual dampness from affecting the new plaster. The wall can then be rendered with a sand/cement mortar, containing a waterproofer. Premixed, lightweight versions, offering similar advantages to lightweight plasters, are obtainable, but **they must be allowed to dry** for about 2 days for shrinkage to occur, otherwise cracking may ensue. A class C plaster is then used for the 'set' (final) coat – see Chapter 6: Plasters. Alternatively, gypsum-based 'renovation' plasters are available; these are designed to dry reasonably quickly in spite of there being some residual moisture in the wall. Whatever system is employed, it is important to deal with the problem at source; that is, to overcome the original cause of the dampness.

4.9 Recycling masonry

Masonry forms a very substantial part of demolition waste and therefore offers considerable opportunities for recycling. Used masonry should therefore be seen as a valuable source of materials for new building, apart from the saving of landfill space and taxes it could afford. Unfortunately, existing buildings are rarely viewed as a starting point for new building, and reclamation prior to demolition is often minimal. Either, the existing building should be viewed as a potential source of new material for its replacement, or, for the longer term, reclaimed materials should be stockpiled for future reuse. A number of materials can be reused with minimal preparation.

Bricks

The potential for reuse of bricks depends largely upon the mortar used to bond the old masonry. In most buildings prior to the 1950s, lime mortars were used. Much of this falls away during demolition so that high reuseage rates can easily be achieved. Superficial damage, for example, to arrises, is often incurred during the process and the subsequent bricklaying operation will therefore involve somewhat more skill. However, very satisfactory results can be achieved provided cost allowance is made for the additional care required. More recent cement-based mortars often bond too strongly to the brick and therefore greatly reduce reuse prospects. Such masonry is commonly recycled as hardcore, though in most cases demand for this is limited and much material is regarded as waste. For the recycling of such waste, see the recycling section of Chapter 3. In view of the high-energy input of bricks, perhaps the recycling issue should be given greater consideration when designing new buildings. There are clearly long-term advantages in the use of lime mortars in buildings of appropriate design.

Slates and tiles

Many of these products have excellent durability. The reuse of slates is already common practice in view of their high second-hand value. There is, however, scope for extending this practice to a wider range of roofing materials and for designers to envisage use of, for example, used clay or concrete tiles, with a possibly weathered appearance, rather than the traditional approach of the clean 'new look' for new buildings. It is known that good-quality hand-made clay tiles in particular often withstand weathering for many generations. This is also a matter of public perception.

Natural stones

These are a low-energy resource, especially as older stone is likely to have been prepared by hand rather than machine. A steady, if small, market already exists, particularly in the northern UK, where nineteenth-century commercial and industrial building was often in stone. Reclaimed stone can be quite difficult to use for new buildings since design must relate to available stocks – especially quantities and sizes.

Experiment 4.1 Determination of the effect of pore diameter on capillary attraction and the magnitude of the surface tension of water

Apparatus
Approximately 200 mm lengths of capillary tubes in a range of bore diameters, for example, 0.2, 0.5, 1, 2 mm; beaker; clamp stand; acetone; distilled water; travelling microscope or accurate weighing balance.

Procedure
Clean the capillary tubes thoroughly with acetone, rinse with distilled water and dry. Do not touch either end of the tube once cleaned. Place each tube in turn in a beaker of distilled water filled to the brim and fixed in a vertical position with a clamp stand. Measure the capillary rise to the nearest millimetre. Repeat for the other tubes. The capillary rise should increase progressively as the tube diameter decreases. If it does not, reclean the tubes and repeat.

Measure the tube diameter with a travelling microscope (or by accurately weighing the tube dry and then with the bore filled with water).

$$\text{diameter (mm)} = 1000 \sqrt{\frac{\text{mass increase (g)} \times 4}{\pi \times \text{length of bore (mm)}}}$$

(Alternatively, bore diameters may be given.)

Plot the capillary rise graphically (y axis) against bore diameter (x axis). Notice the trend in the capillary rise for very small bore diameters. Pores in building materials may be as small as 10 nm (10 nm = 0.01 μm = 10^{-8} m). The surface tension (T) of water can be calculated from the formula:

$$T = \frac{R\rho g h}{2}$$

where R = radius of bore

ρ = density of water (1000 kg/m^3)

h = capillary rise

g = acceleration due to gravity.

Experiment 4.2 Measurement of bulk density, solid density, porosity and degree of saturation of porous building materials

Apparatus
Oven; beaker; relative density bottle; balance accurate to 0.1 g and a further balance accurate to 0.0001 g; clamp stand; cling film; fine thread; mortar and pestle; 150 μm sieve; funnel.

Specimen preparation
Pieces of brick, concrete or insulation block of size about 50 × 50 × 50 mm are suitable. It is helpful if materials which float on water are cut accurately so that the bulk volume can be determined by dimensional measurement.

Procedure

Dry the specimens thoroughly in an oven at 110 °C and allow to cool in air.

To measure bulk density (D_B)

Weigh the specimen to 0.1 g accuracy. Find bulk volumes as follows:

(a) Specimens which float in water – for example, aerated concrete blocks. Obtain the volume by dimensional measurement:

$$\text{vol. in litres} = \frac{\text{vol. in mm}^3}{10^6}$$

(b) Specimens which sink in water. Wrap carefully in cling film. Fill a beaker with sufficient water to enable the specimen to be totally immersed without causing the beaker to overflow. Place the beaker in a top-loading balance and note the mass in g (0.1 g accuracy). Wrap the specimen carefully in cling film, making sure there are no trapped air pockets, and attach a thread. Suspend, using a clamp stand, in the beaker of water, avoiding contact with sides or base. Note the new balance reading in grams.

$$\text{bulk volume (ml)} = (\text{new reading} - \text{old reading})$$

Hence find the bulk density (D_B)

$$D_B = \frac{\text{mass (g)}}{\text{bulk volume } in \text{ litres}} \quad \text{g/l} \quad \text{or} \quad \text{kg/m}^3$$

To measure solid density (D_S)

Pulverise about 20 g of the material, using a mortar and pestle, until it passes a 150 μm sieve. Weigh out about 10 g of the material, using the sensitive balance (mass M_1 g) and transfer, using a funnel, to the relative density bottle, avoiding spillage. Add one or two drops of soap solution to a beaker of distilled water, stir and half fill the relative density bottle, using the funnel. Swirl gently until *all* solid material has sunk and entrapped air is removed. Fill to overflowing, insert the stopper and dry the outside of the relative density bottle. Weigh (M_2 g). Empty and clean the bottle, refill with distilled water, insert the stopper, dry and weigh (M_3 g).

The solid volume of the material is

$$M_1 - (M_2 - M_3) \quad \text{ml}$$

The solid density (D_S) of the material is

$$D_S = \frac{M_1}{M_1 - (M_2 - M_3)} \quad \text{g/ml}$$

(multiply by 1000 to change to kg/m^3).

Use the following formula to calculate the porosity (P)

$$D_B = D_S \left(\frac{1 - P}{100} \right)$$

To find the water absorption and the percentage pores filled on saturation

The sample used for the bulk density determination (oven dry mass W_1 g) can also be used for this determination. To obtain an indication of the degree of saturation obtained by short exposure (say to rain), immerse in water for a period of about 10 mins, and weigh again (W_2 g).

The percentage water absorption is

$$\frac{W_2 - W_1}{W_1} \times 100$$

The degree of saturation of the sample is found as follows:

$$\text{bulk volume} = \frac{W_1}{D_B} \quad \text{litres} \quad (D_B \text{ in g/l})$$

$$\text{solid volume} = \frac{W_1}{D_S} \quad \text{litres} \quad (D_S \text{ in g/l})$$

Volume of pores equals bulk volume minus solid volume; that is,

$$W_1 \left[\frac{1}{D_B} - \frac{1}{D_S} \right] \text{litres} \quad \text{or} \quad 1000\, W_1 \left[\frac{1}{D_B} - \frac{1}{D_S} \right] \text{ml}$$

Volume of water absorbed = ($W_2 - W_1$) ml. Hence percentage of pores filled (degree of saturation) is

$$\frac{W_2 - W_1}{1000\, W_1 \left[\dfrac{1}{D_B} - \dfrac{1}{D_S} \right]} \times 100$$

This experiment should be repeated, boiling the sample for some time to determine the total proportion of pores that is accessible to water. The performance of different bricks or block types should also be compared.

Experiment 4.3 Measurement of water absorption of different types of mortar

Apparatus

70 mm mortar cubes; gauging trowel; weighing balance; plasticiser; curing tank; oven.

Procedure

Make up samples of mortar using the following quantities of cement and builder's sand:

Ratio cement/sand	1 : 3	1 : 6	1 : 9	1 : 12
Cement (g)	250	150	100	80
Sand (g)	750	900	900	960

Mix each into a workable mortar with water which has been plasticised either with a mortar plasticiser or a few drops of washing up liquid. The consistence of each should be the same.

Fill 70 mm mortar cubes with each, identify and cover.

Demould after one day and cure underwater until one week old.

Oven-dry the cubes. Weigh each cube and record the masses. Immerse in water and leave for five min. Record masses again. Re-immerse and obtain masses after 10, 20 and 60 min. Calculate the water absorption of each, as a percentage, at each stage.

$$\text{water absorption} = \frac{\text{water absorbed}}{\text{dry mass}} \times 100$$

Plot water absorption against time. Comment on the relation for the different mortar types and give implications regarding the use of these mortars.

Experiment 4.4 Determination of the compressive strength of clay bricks

This experiment is much more valuable if 10 bricks of a given type are tested, since the variability of the bricks can then also be ascertained. If this is not possible, a smaller number of, say, three bricks might be tested to give some indication of the ranges.

Apparatus

Gauging trowel, four 75 mm cube moulds for mortar, curing tank; compression tester; plywood sheets of size 225 × 112 mm and thickness 4 mm – two per brick tested, assuming standard metric bricks.

Preparation of bricks

All bricks should first be soaked for 24 h and then removed and allowed to drain for 5 mins. If bricks have frogs which are to be laid frog-up, the frogs must be filled with mortar prior to testing. The mortar should comprise a 1 : 1.5, cement/sand mix with sufficient water for satisfactory workability. (To achieve the required 3–7 day strength with very fine sands, it may be necessary to use a 1 : 1 cement/sand mix). Fill the frogs with mortar and, after allowing 2–4 h for stiffening, trowel as smooth and flat as possible. Make four 75 mm cubes from the mortar. Store the bricks and cubes under damp sacking for 24 h, demould cubes and transfer both bricks and cubes to a curing tank. Bricks may be tested when the mortar strength is between 28 and 42 N/mm² (usually achieved in 3–7 days).

Procedure

Remove the bricks from the water and wipe away excess water and any grit. Label the bricks and record the dimensions of the smaller bed face of each brick. Mount centrally in the compression machine with plywood sheets above and below. The bricks should be mounted in the same orientation as they are intended to be used in service (normally frog-up). Apply the load at 15 N/mm² per min (approximately 330 kN per min for standard bricks) until maximum load is reached. Record this load and calculate the stress at failure, using the dimensions of the smaller bed face. Repeat for the other bricks and find:

(a) the average strength;
(b) the range of strength.

Hence classify the bricks (see p. 124).

Experiment 4.5   Determination of the water absorption of clay bricks

To obtain a meaningful result, the BS 3921 boiling or vacuum test should be carried out. The vacuum test is much quicker than the boiling test but requires additional apparatus so that the latter test will be described here for simplicity. It is suggested that two brick types should be investigated – for example, Flettons and engineering bricks.

Apparatus
Weighing balance capable of 0.1 g accuracy; tank of capacity sufficient to heat 10 specimens; drying oven.

Procedure
Ten bricks should, ideally, be used, though they may be sawn into half or quarter sections to save space in the heating tank. Dry the bricks for 2 days at 100 °C and allow to cool to room temperature. Label and weigh the bricks and record the masses. Place in the water tank to allow water to circulate freely. Heat to boiling and maintain the temperature of 100 °C for 5 h. Allow to cool *still submerged in the tank* for 16–19 h. Remove the bricks, wipe off surface moisture and weigh immediately.

Calculate the water absorption (see p. 124) and compare with the limits for engineering bricks (Table 4.2).

Experiment 4.6   Measurement of efflorescence in clay bricks

(Note that this is no longer a BS test although it serves to illustrate the principles of the problem.)

In common with other tests, 10 specimens should be used, since results often vary considerably from brick to brick.

Apparatus
Ten wide-mouthed bottles or flasks; large flat tray.

Procedure
The test should be carried out in a warm, well-ventilated room. Wrap the bricks in polyethylene sheet, leaving only the face which is to be exposed in the brickwork uncovered. Tie with a rubber band or string (Fig. 4.18). Fill the first bottle with distilled water and place a brick over the mouth so that it completely covers the opening. Invert both brick and bottle while maintaining them in close contact and stand in the tray. Repeat for the other bricks. Leave for a few days, replacing the distilled water if it is completely absorbed within 24 h. When the bricks have dried,

Fig. 4.18 *Apparatus for efflorescence test.*

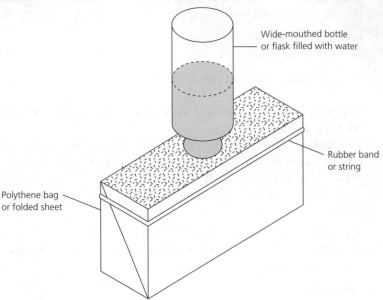

Wide-mouthed bottle
or flask filled with water

Rubber band
or string

Polythene bag
or folded sheet

add more water and repeat the process. After drying for the second time, examine the bricks for efflorescence and classify, using the following criteria:

No perceptible deposit of efflorescence	Nil
Not more than 10 per cent of area covered with thin deposit of salts but unaccompanied by powdering or flaking of the surface	Slight
A heavier deposit than slight, covering up to 50 per cent of the area of the face but unaccompanied by powdering or flaking of the surface	Moderate
A deposit of salts covering more than 50 per cent of the area of the face and/or powdering or flaking of the surface	Heavy

Experiment 4.7 Measurement of frost damage to clay bricks

Frost usually takes a number of years to result in visible damage to brickwork in practice, so that a laboratory test must attempt to accelerate the process. A fully realistic test must also include the effect of mortar joints. Panels comprising 30 or more bricks would be ideal but would require a very large refrigerated enclosure. This experiment is designed to give some indication of the likely properties of a brick, using only a small domestic freezer, though a period of some weeks will still be necessary for its execution.

Apparatus

Shallow containers; domestic freezer (upright type is more convenient); compression test machine.

Procedure

Ideally 10 or more bricks of each type should be subjected to freeze/thaw cycles, otherwise variability between individual bricks may overshadow the effects of frost, especially in naturally durable bricks.

Take a batch of 20 bricks of the type required and place half underwater at room temperature for the duration of the test. Place the other 10 in shallow trays of water in the freezer. Subject the bricks to temperature cycles in the range 20 to −20 °C, keeping continuously saturated. With an upright freezer one cycle per day should be possible – for example, switch on at 17.00 hours and by 09.00 hours the next day the temperature of −20 °C should have been reached. At this point, switch off the freezer and leave the door open so that the bricks thaw *fully* – a period of 8 h should suffice. (The thawing cycle may have to be lengthened for chest freezers, which are slower to thaw unless the bricks are removed.) Repeat the procedure, topping up trays with water as necessary.

Inspect every 10 cycles for signs of visible damage. After 50 cycles, thaw the bricks and test in a compression machine (see Experiment 4.4). Measure the average strength of the bricks that were subject to freezing and express as a percentage of that of the remaining bricks.

Note that this is a severe test and, although no clear criterion exists for frost resistance, bricks might be assumed to have a high degree of frost resistance if their strength after 50 cycles has not decreased by more than 10 per cent and there is no visible flaking.

Experiment 4.8 Measurement of crystallisation damage in clay bricks

Crystallisation damage, in common with frost damage, usually occurs progressively over a number of years and this simple test is therefore designed to accelerate the process, though it may still take several weeks to complete.

Apparatus

Large tray in which 10 or more bricks can be immersed; plastic rack or grid on which bricks can be dried; compression machine; 5 per cent solution of magnesium sulphate.

Procedure

To accelerate drying, the test should be carried out in a warm, well-ventilated room. Place 10 bricks in the solution so that they are completely covered. Place another 10 bricks of the same type underwater for the duration of the test. After a few hours, remove the bricks from the solution and place on a rack or stand which allows them to drain freely. Allow to dry completely – this should be easy to ascertain, since dry bricks are usually a lighter colour. The process may take several days, depending on the type of brick and drying conditions.

Repeat the procedure until at least 20 wetting and drying cycles have been applied, topping up the salt solution as necessary.

Finally, rinse in ordinary water and observe any visible effects on the bricks. Test the 10 bricks in a compression machine, as described in Experiment 4.4. Test also the 'control' bricks and express the average strength of the bricks subject to the salt solution as a percentage of that of the others.

Crystallisation may be assumed to have had a significant effect on the bricks if there is visible damage or if the strength has decreased by more than 10 per cent.

SELF-ASSESSMENT QUESTIONS

1. Describe the production of:
   (a) Fletton bricks;
   (b) engineering bricks.
   Give characteristic properties of both types of brick and two common applications of each.

2. Explain the meaning of the term 'sintering' as applied to clay bricks. Indicate the effects on performance when bricks are
   (a) overfired;
   (b) underfired.

3. Compare and contrast the moisture movement properties of clay bricks and calcium/silicate (sand/lime) bricks.

4. A clay brick of length 220 mm and width 106 mm failed in compression at a load of 1830 kN. After 5 h boiling in water, the mass increased from 2.94 to 3.05 kg. Use Table 4.2 to classify the brick, assuming it was typical of its type.

5. Explain what is meant by 'crystallisation damage' in clay brickwork. Give situations in which it is especially likely to occur. Give three steps that may be taken to minimise the risk of damage.

6. State the cause of sulphate attack in brickwork. Give three precautions by which the likelihood of such attack can be reduced.

7. State what is meant by 'efflorescence' and indicate when and where it is most likely to be seen. Explain why efflorescence is unlikely in sand/lime brickwork.

8. Give the main reasons for the current widespread use of autoclaved aerated concrete blocks. Indicate why it is important that these blocks are not saturated prior to use.

9. Indicate four performance requirements of mortars for masonry. State, with reasons, proportions for a mortar mix that would be suitable for
   (a) a wall for a domestic property to be built using 'stock' bricks in winter weather;
   (b) an internal wall using autoclaved aerated concrete blocks.

10. Suggest how a wall for a domestic building could be constructed using natural stone, indicating how adequate structural performance and thermal insulation could be obtained.

11. Explain why two-coat work is essential for a render coating to a solid wall for domestic construction. Suggest why fine sands would not be suitable for the purpose.

12. State the meaning of the following terms:
    (a) cohesion;
    (b) adhesion.
    Hence explain the meaning of the terms 'surface tension' and 'capillarity'. Give simple illustrations of the behaviour of water which results from these two phenomena.

13. Describe how the degree of attraction of water into water pores varies with pore diameter. Hence indicate whether course or fine-pored materials tend to be quicker drying after being saturated.

14. A lightweight block is found to have a solid density of 2400 kg/m^3 and a bulk density of 800 kg/m^3. Calculate its porosity.

15. A common brick has a bulk density of 1300 kg/m^3. If the porosity is 45 per cent, calculate the solid density of the material.

16. Explain the meaning of the term 'suction' in bricks. Give two advantages that result from bricks having some suction.

17. Give four possible cases of damp patches on internal plaster, indicating areas most likely to be affected by each type.

REFERENCES

Building Research Establishment Digest 245: *Rising damp in walls: diagnosis and treatment*, 1986
Building Research Establishment Digest 362: *Building mortar*, 1991

STANDARDS

BS 187: 1978: *Calcium silicate (sand lime and flint lime bricks)*
BS 1199 and 1200: 1976 (1996): *Building sands from natural sources*
BS 3921: 1985 (1995): *Clay bricks*
BS 5618: 1985 (1996): *Code of practice for thermal insulation of cavity walls (with masonry or concrete inner and outer leaves) by filling with urea formaldehyde (UF) foam systems*

BS 5628: *Code of practice for use of masonry*
    Part 1: 1992: *Structural use of unreinforced masonry*
    Part 3: 1985: *Materials and components, design and workmanship*
BS 6073: *Precast concrete masonry units*
    Part 1: 1981: *Specification for precast concrete masonry units*
    Part 2: 1981: *Methods for specifying precast concrete masonry units*

5 Glass in construction

Chapter summary

- Strength of glass.
- Safety with glass – toughened, laminated and wired.
- Thermal performance.
- Decorative use of glass.

5.1 Introduction

The term 'glass' refers to any material which is largely amorphous (non-crystalline) in its solid state. Effectively, 'glasses' resemble liquids in that they have no ordered molecular structure. This confers very important properties on the product. Quite high strengths are at least theoretically obtainable because there are no weaknesses associated with the grain (crystal) boundaries found in compounds which crystallise. Perhaps more important still for the most widespread use of glasses in building (glazing), is that they tend to be transparent, again because there are no crystal boundaries to intercept or reflect light. Each sheet of glass could be regarded as a single very large molecule. Many plastics lacking an ordered molecular structure are also referred to as 'glasses'. They will be dealt with in Chapter 8, but our attention is focused in this chapter upon glass products based on silica. It is important to appreciate that silica is one of the commonest minerals in the earth's crust since it comprises oxygen, the commonest element, and silicon, the next most common element. Hence supplies of the raw materials could be considered to be virtually inexhaustible.

Glazing, which forms a very important part of most buildings, has the following attractions:

- It admits daylight while providing a weatherproof barrier.
- Daylight is known to have psychological benefits for people compared with artificial light sources.
- Plain glazing provides the interest of an external view.
- Glazing provides an increased feeling of spaciousness in a room of a given size.

Table 5.1 *Environmental considerations: glass*

Consideration	Assessment
Raw material availability	Materials in abundant supply
Extraction	No specific problems in volumes currently required
Energy used in manufacture	Furnace required −1100 °C
Health/safety hazards	Principal danger is lacerations due to sharp edges
Recyclability potential	Excellent
Disposal	Does not break down readily in the ground

Glass is the most traditional material for glazing, offering the following benefits:

- Glass is quite cheap to produce.
- Glass is relatively hard and generally scratch resistant.
- Glass is quite inert and largely unaffected by exposure to normal atmospheres.
- The material is easy to cut.
- It has excellent optical properties.

Possible problem areas include:

- There are serious injury risks due to the sharp edges exposed by cutting or breakage.
- Ordinary glass has poor thermal insulation properties leading to high heat gain in summer and high heat loss in winter.

Environmental considerations (Table 5.1) include:

- Glass is very suitable for recycling – in fact used glass is an essential part of the glass-making process since it has a relatively low melting point.
- Conversely, if not recycled, glass does not readily break down over time and can be a hazard.
- High temperatures are required for its manufacture (though in the quantities used in building this should not constitute a serious environmental problem).

5.2 Manufacture

Glass consists in essence of silica (silicon oxide) though this material alone is unsuited to the manufacture of glass because it has a high melting point and would tend to crystallise on cooling. Other ingredients are therefore added: chiefly sodium carbonate to reduce the melting point, and calcium carbonate to reduce the water solubility of the product. One advantage of amorphous materials is that small amounts of other compounds can easily be added to modify properties while retaining the essential ingredients of glass, as outlined above. Examples are

- Borosilicates to reduce thermal movement.
- Lead to improve surface lustre.

In addition, the surface of glass can be modified by coatings applied and fused into the surface in order to control light reflection, for example.

The traditional way of producing glass for ordinary glazing involved drawing off liquid material from a tank of molten glass using a metal 'bait', the sheet being engaged in rollers to prevent waisting. This process produces a slight ripple in the glass, which can be seen by moving slightly relative to a window while observing an external object or building. The image 'moves' slightly as the line of sight passes the varying thickness of the glass. More recently the process has largely been replaced by the float glass process in which the hot, soft material is passed over a bed of molten tin. This greatly reduces thickness variations and leads to distortion-free vision.

Glass can also be produced by rolling, various patterns being obtainable. This produces a textured (obscured) glass which allows light transmission while providing some privacy, for example, in glazed screens. Wire can be incorporated in the rolling process and this greatly improves performance in fire and reduces the risk of injury if failure occurs in overhead or low-level glazing.

5.3 Strength of glass

It might be expected that because each sheet of glass is effectively one very large molecule, glass should be very strong, reflecting the strength of the primary (covalent bonding) which it largely comprises. The strength of the material is, however, greatly reduced by the effect of minute surface defects which are invariably present. These concentrate the stress around them. Hence, if a sheet contains a minute scratch the stress at the tip of the scratch may be hundreds of times the average stress in the material. The situation is represented in Fig. 5.1, showing stress 'lines'. The closer the lines are

Fig. 5.1 *Concentration of tensile stress caused by microcracks in glass surface.*

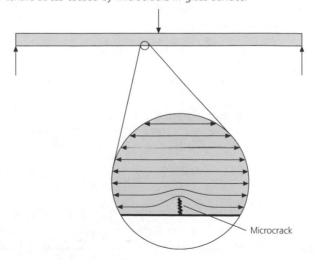

Microcrack

together, the higher the stress. New glass may have a flexural strength of more than 100 N/mm^2, which, although much below a defect-free level of performance, is much better than, say, concrete, which in flexure rarely sustains more than about 5 N/mm^2.

On ageing, glass tends to become scratched or subject to surface damage by pollution, these additional defects further reducing the strength. Such defects also make glass cutting more difficult. A glass cutter introduces a continuous 'defect' by means of a hardened wheel. If additional defects are present from ageing there is a higher risk that, on cutting, the crack may propagate along an existing defect, rather than along the 'defect' produced by the cutter. Unlike metals, glass does not have the ability to flow and is therefore very sensitive to the effects of flaws or concentrated stresses in the material. For this reason glass should never be allowed to contact hard metals such as steel nails or fixing bolts which can generate very high local stresses and initiate a crack. 'Ideal' fixing materials are rubber gaskets, putties (though these can become very hard on ageing, especially if not protected by paint) and soft metals such as lead, as in leaded lights.

5.4 Safety with glass (BS 6206)

Each year, hundreds of injuries, sometimes fatal, are produced by glass splinters or 'shards'. They result from the very sharp edges formed when glass breaks. (The fine particles of many abrasive papers are based on this property.) The problem can be greatly reduced by one of the processes detailed below.

Toughening

The surface of the glass is subjected to cool air jets while in the hot state. These surface layers cool relatively quickly while the inner core is still soft. When the inner layers subsequently contract they impart a compressive stress to the already cool surfaces. The situation is shown in Fig. 5.2. The surface compressive stress is balanced by a small tensile stress though the bulk of the glass. This considerably improves the performance of the glass since, in flexure:

1. The (convex) surface of the glass in flexure would normally tend to be subject to the highest tensile stress.
2. Failure is usually initiated at surface flaws, as explained above.

Fig. 5.2 *Stress distribution through a sheet of toughened glass. Compression at the surfaces closes the cracks. The tensile stresses at the core are not a problem because there are no cracks at the centre.*

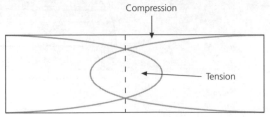

Compression

Tension

Toughened glass has therefore greater bending strength than plain glass, and as its impact strength in particular is about five times greater, the thickness required for a particular application can be reduced. In addition to the higher performance of toughened glass, when fracture takes place it tends to occur throughout the sheet as the internal balance of stress is upset. The glass therefore crazes extensively and on complete failure shatters into large numbers of quite small pieces with a comparatively small hazard to safety. Note, however, the following:

1. Toughened glass cannot be cut. This destroys the stress balance through the material and would cause shattering. It is most important to order the material in the correct size, allowing for tolerances, especially if large numbers of sheets are required.
2. Visibility is lost when the glass fails and if disintegration occurs there is loss of weather protection and security.

Laminated glass

This comprises two more layers of plain glass bonded by sheets of a clear plastic called polyisobutyrate. The plastic absorbs energy during impact so that the toughness of the material is greatly increased. Unlike toughened glass, the material can be cut (though cutting must be done in stages, first cutting the glass on each side and then softening/cutting the plastic film).

The major advantage of laminated glass is that 'failure' does not lead to disintegration, and visibility, weatherproofing and security are normally retained. The material does not, however, have the same strength properties of toughened glass since it comprises only normal strength sheets.

Wired glass

This is the most traditional method of improving structural performance of glass, the wire holding together the material on cracking. Although it does not have the resistance to failure of the above materials it has the possible advantage that the wires can be clearly seen. It therefore has some deterrent effect when vandalism is a possibility.

Special glasses should be used in overhead glazing, in low-level glazing, in doors and in large glazed areas. When considering the appropriate type of glass and the design of glazing in a given situation, the following points should be considered:

- Loadings such as wind need to be withstood.
- Need for security or weatherproofing on failure.
- Emergency evacuation, for example, in fire – especially if windows are non-openable.
- Risks of large plain glazed areas to the young and elderly who may not see them.

BS 6206 (soon to be replaced by the corresponding Eurostandard) specifies that safety glass should either not break or 'break safely' under standard impacts. The severity of these impacts was originally designed around the likely kinetic energy of a running child. For example, a child weighing 50 kg running at a speed of 5 m/s would have a kinetic energy of

Fig. 5.3 *Particle count requirement for toughened glass. The count in a 50 mm square should be at least 40 in the coarsest fracture area (part pieces in the square count as 0.5). This piece easily passes the test. (Adapted from BS 6206: 1981)*

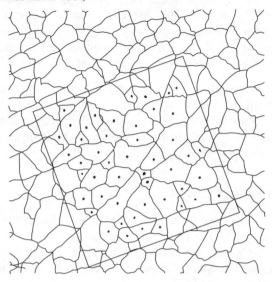

$$\frac{mv^2}{2} = \frac{50 \times 5^2}{2} = 625 \text{ Joules}$$

A 45 kg mass is used for the test (to simulate the weight of a child) dropped through heights corresponding to impact energies of 135, 202 and 538 J (see Fig. 1.8). These are lower than the value given above since actual impact energies are found to be less than the theoretical figures. Classes A, B and C are assigned accordingly, class A being the highest rating for the 538 J test.

The term 'break safely' is interpreted differently for toughened glass and other safety glasses (which could include laminated glass, wired glass or glass bonded to a plastic film):

- Toughened glass either must not break or if it does, the size of the 10 largest broken fragments should not exceed the mass equivalent of 6500 mm^2 of the original sheet. This limits the risk of injury from shards of glass. The glass is also subject to a separate shatter test, the number of particles in a 50 by 50 mm square at in the region of coarsest fracture being required to be at least 40 (Fig. 5.3).
- In the case of laminated glass, wired glass or glass bonded to a plastic film fractures may occur but there should be no openings through which a 76 mm diameter sphere could be passed. In addition there are particle size restrictions on any detached pieces.

Safety plastics are included in the standard with separate requirements for fracture openings and fragment sizes.

All safety glass should be indelibly marked, including the impact rating, A, B or C, and in view of the safety implications, specifications should require third party certification such as the BS Kitemark as independent evidence of conformity.

5.5 Thermal performance

Glass for glazing has considerable architectural possibilities in addition to the indisputable advantages outlined in the introduction to this chapter. Nevertheless, glass, especially in the form of large glazed areas, can pose considerable thermal problems for occupants in summer and winter.

Summer

The almost complete transparency of plain glass to visible light forms the basis of its use for glazing purposes. The sun's rays comprise visible light, some ultraviolet and some near infrared radiation in a continuous wavelength band from about 0.3 to 3 μm, most of the energy being in the visible/near infrared region. Plain glass transmits most of this radiation, as can be experienced when sitting in sunlight behind glass. This radiation is absorbed by, and heats, internal surfaces which then re-radiate energy. The re-radiated energy is of a higher wavelength than the original, in the range 3–30 μm. (Note that this radiation is completely invisible and not directly related to the colour of a surface which is based upon the absorption characteristics of the surface pigmentation.) The re-radiated longer wavelength energy is largely absorbed by the glass (Fig. 5.4). The glass becomes warm but loses some heat to the external air. It therefore effectively traps the sun's rays – the 'greenhouse' effect. For this reason large glazed areas can be most uncomfortable in summer. The problem is most acute in south-west facing aspects later in the day when sun shines through vertical glazing into rooms already warmed by occupation and the heat of the day. Substantial gains can occur from spring to autumn, the worst effect often being early and late summer when the sun is not at its full height in the sky. Possible remedies include:

Fig. 5.4 *Greenhouse effect in which the sun's rays are trapped in an enclosure. The light arrows represent the sun's rays (UV) and the dark arrows internally emitted radiation (IR). The size of arrow represents the amount of radiation.*

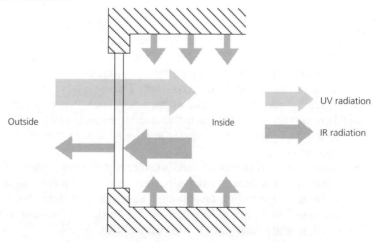

Fig. 5.5 *Illustration of reflective and tinted solar control glass. Reflections can be seen in glass at the left and right sides of the building. The central glazed area is tinted, giving a view of the interior. (Sun Alliance Building, Basingstoke.)*

1. Use tinted, heat-absorbent types of glass. These are available in a number of shades including grey and green, the glass itself absorbing as much as 50 per cent of the heat (and a similar proportion of light, though some heat will be re-radiated through the glass. The amount of heat absorbed depends on the glass thickness and care is necessary in design with such sheets to prevent excessive temperature rise in the glass, with consequent overstressing, especially if glazing beads shield edges from the heat, producing a cool glass perimeter. Tinted glass permits a view of the interior of the building.

2. Heat-reflecting glass can be achieved by thin metallic surface coatings, usually applied to the inner face of the outer glass sheet in sealed double glazing, for protection, though they can also be used for single glazing. As much as 40 per cent of the heat, and a similar proportion of light, are reflected and, in this case, there is less heat absorption of the glass itself. These glasses have a metallic tint which may, typically, give a mirror-like appearance when viewed externally (Fig. 5.5) while there is a bronze tint to the external environment, when viewed from the inside. (As with tinted glass, the occupants tend only to be aware of tints when a window is opened.)

3. Blinds or louvres can be used to intercept the sun's rays. These are much more effective when situated externally since they prevent the heat reaching the enclosure. Current practice often favours this technique and some types can be controlled to combine maximum effectiveness at any time of the year with minimum loss of daylight and visual contact with the exterior.

Winter

It has already been commented that glass absorbs internal heat and does not therefore permit the sun's absorbed energy to disperse effectively in summer. In winter the same property leads to increased heat losses, the sun's contribution to warming usually being negligible. Glass absorbs radiant heat from the room and transmits it readily to the exterior when the outside air temperature is low. Ordinary glass has a high thermal conductivity and therefore has little inherent ability to insulate.

For example, a surface temperature drop of only 10 °C across a 1 metre square sheet of glass of thickness 4 mm would result in a heat flow of almost 3000 watts – a high heat loss for such a small area.

The main resistance of glazing to heat comes in fact from air layers at each surface which impedes the convected heat flow from the warm interior air to the cooler glass surface. Double glazing operates by providing twice as many of these glass/air interfaces, though a cavity width of 20 mm is the optimum for obtaining full advantage of the insulating qualities of the air trapped between panes. Sealing of the cavity is always recommended to prevent dirt and condensation problems and a further 10 per cent reduction in U value can be obtained by filling the cavity with an inert gas such as argon. Yet another possibility is to reduce the absorption of the glass to internal radiant heat by low emissivity coatings. Where necessary, effective heat retention in winter months can be combined with solar control in summer months by use of double-glazing systems incorporating solar control glass.

Building Regulations in relation to glazing heat losses

To save energy in dwellings, current Building Regulations restrict the size of glazed areas which normally have much higher losses than other elements of construction.

The allowable glazing heat loss in any one enclosure is the amount which would be lost from a single-glazed area of 12 per cent of the total perimeter wall area of that enclosure.

This figure represents a typical average glazed area of traditional style housing in the UK, though larger glazed areas could be incorporated as follows:

- Use double glazing. This would permit about double the glazed area to be used on the basis of the value of 2.8 W/m² °C prescribed for double glazing, as against 5.7 W/m² °C for single glazing.
- Use treble glazing (U value 1.9 W/m² °C or less). This would permit three times the 12 per cent figure indicated above.
- If the external walls have a U value less than 0.6 W/m² °C, then additional glazing can be employed provided the total heat loss does not exceed that corresponding to an external U value of 0.6 W/m² °C and 12 per cent of single glazing.

The Building Regulations also allow other glazing systems, provided the architect/services engineer can demonstrate compliance with the basic energy requirements as outlined above. It is likely that, in the near future, heat insulation regulations will be tightened to reduce further heat loss in domestic buildings.

Fig. 5.6 *Production of a leaded stained-glass light for a door.*

5.6 Decorative use of glass

Glass has for centuries been used for decorative as well as light transmission purposes. By use of obscured and/or stained glass, glazing can provide highly attractive displays within the window itself and interesting internal lighting effects. Metallic oxides can be incorporated into the glass at firing stage and become a part of the material, so colours are completely permanent. Examples of colours that are easily obtainable are:

Red   copper
Blue   cobalt
Green   iron
Yellow   manganese

Alternatively, colours can be painted onto plain glass and sealed by a resin or heat treated to improve stability. These colours do not have the purity of the fused metallic oxides.

Traditional use involves bonding together of small sheets of glass by lead 'cams', usually of H section, into which carefully cut pieces are slotted. A typical procedure for a stained door glass is shown in Fig. 5.6. A full-size design is drawn and carefully cut pieces of glass are assembled into the cams using a simple wooden frame as a guide. On completion, the upper joints are soldered, the now rigid unit is turned over and the other side is soldered. Finally, a special lead light cement is worked into the joints to make them waterproof. The cement stiffens over time and the final unit is quite rigid.

Leaded light glazing deteriorates slowly, especially in polluted atmospheres or where vibration due to impact or traffic is present, but the glass itself should last for long periods. Maintenance may take the form of renewing the lead when necessary, but where minimum intrusion is required in older examples, a sheet of clear glass can be applied on the exterior side to give protection from further deterioration.

SELF-ASSESSMENT QUESTIONS

1. Give three advantages and three disadvantages of glass for glazing purposes, as compared with clear plastics products for the same purpose.

2. Explain the aspects of glass which limit its structural properties and indicate how properties can be improved by
   (a) toughening and
   (b) laminating.

3. Explain why large glazed areas can lead to comfort problems in both summer and winter. Indicate in each case possible ways of improving performance.

STANDARD

BS 6206: 1981 (1994): *Specification for impact performance requirements for flat safety glasses and safety plastics for use in buildings*

6 Plasters

Chapter summary

- Types of gypsum plaster.
- Lightweight aggregates.
- Plaster systems.
- Plaster for damp applications.
- Plasterboards.
- Common defects.

6.1 Introduction

Plasters may be defined as materials designed to provide a durable, flat, smooth, easily decorated finish to internal walls or ceilings. Plasters were traditionally based on lime or cement, but in the last 30 years gypsum has become the most important binder. Gypsum plasters have the following advantages:

- Their setting time can be controlled according to function.
- There is minimal waiting time between successive coats.
- Various surface textures and surface hardnesses are obtainable according to function.
- Unlike cement-based plasters, they are non-shrinking (provided plastering technique is correct).
- They have excellent fire resistance. Set gypsum plaster contains 21 per cent water of crystallisation which absorbs considerable heat, minimising the rate of temperature rise in and behind the plaster. Calcined (heated) gypsum also acts as an efficient insulating barrier in fire.

Note that:

- Since gypsum is slightly soluble in water, gypsum plasters are not suitable for exterior uses unless very effective permanent protection is provided.
- Gypsum plasters are slightly acidic and will causes quite rapid rust staining of embedded steel during setting/drying, or if they become damp in use.

Table 6.1 *Environmental considerations: plasters*

Consideration	Assessment
Raw material availability	Abundant
Extraction	No specific problems. Much extraction is carried out underground, avoiding damage to landscape
Energy used in manufacture	Moderate heat is involved in manufacture (180 °C approx)
Health/safety hazards	There are no specific problems, though, as with any other dust, inhalation (especially at mixing stage) should be avoided
Recyclability	Difficult as it is usually bonded to other materials

- Gypsum plasters have a relatively short shelf life. The commonest class B plasters exhibit a rapid set if use is delayed for more than about two months, depending on storage conditions. Bags are date stamped and should be used in strict rotation, and within this period. Very poor finishes may be obtained if the plaster sets chemically before the normal plastering process has been completed.

Environmental considerations are given in Table 6.1.

6.2 Plaster types

A maximum of three coats may be used, each having its own function:

1. Render coat – to level the background.
2. Floating coat – to produce a flat surface of uniform suction.
3. Finishing coat – to provide a smooth, hard finish.

The first two coats require relatively coarse-textured plasters, applied in thicknesses of up to 20 mm. The functions of these are now largely combined in a single undercoat or 'browning' coat. The finish coat utilises a much finer material of thickness up to 5 mm. Some backgrounds, such as plasterboard, provide a flat surface with uniform suction, so that a single finishing coat may suffice.

6.3 Classes of gypsum plaster (BS 1191)

The raw material is calcium sulphate dihydrate ($CaSO_4 2H_2O$), obtained from mines. During manufacture, some or all of the water is driven off by heat, the nature of the resultant material depending on the heating regime. Small amounts of impurities are usually present and these colour the plaster grey or pink, though they have no other significance.

BS 1191 classifies gypsum plasters as follows:

Class A: hemihydrate $CaSO_4\frac{1}{2}H_2O$ (plaster of Paris)

This is produced by heating to a temperature not in excess of 200 °C. Plaster of Paris sets within 5 to 10 minutes of adding water, which is too rapid for use in ordinary trowel trades. It is nevertheless useful for casting purposes, such as in decorative plasterwork.

Class B: retarded hemihydrate $CaSO_4\frac{1}{2}H_2O$

These are produced from class A plasters by the addition of a suitable set retarder such as keratin. The amount of retarder added depends on function.

Undercoat plasters tend to be slower setting than finish coat plasters to allow time for straightening. These plasters are normally designed to be used with sand in ratios of up to 3 : 1 sand/plaster. Increasing the sand content has the effect of accelerating the set, setting times being in the range 2–3 h.

Finish coat plasters are designed to be used neat or with the addition of up to 25 per cent by weight of hydrated lime, which accelerates the set. Setting times are in the range $1–1\frac{1}{2}$ h.

Class B plasters should not be re-tempered once setting has commenced. A most important application of class B plasters is in premixed plasters containing lightweight aggregates, which are now very widely used. 'Board finish' plasters for plasterboard are also class B.

Class C: anhydrite

Class C plasters are obtained by heating the raw material to a higher temperature than class B plasters, with the result that a proportion of anhydrous calcium sulphate forms, although some hemihydrate also remains. The anhydrous component is so slow to set that an accelerator such as alum is added. The presence of two compounds gives class C plasters a 'double set' – an initial set and then a slow final set, so that the material can be re-tempered with more water up to 30 minutes after adding the water, though it should be applied as soon as possible after mixing and sets completely in $1\frac{1}{2}$ hours. An important current use of class C plasters is a finishing coat on a sand/cement backing (trade name 'Sirapite').

Class D: anhydrite (Keens's cement)

This is harder burnt than class C plaster, resulting in a higher proportion of anhydrite. The plaster set must again be accelerated and, as in class C plasters, stiffening is continuous. The final product has superior strength, smoothness and hardness compared with other types and is used for arrises and surfaces such as squash court walls, where a very durable finish is required. It also provides an ideal base for gloss paints. On account of its strength, it should not be used on soft backings such as plasterboard or fibreboard.

Note
All plasters set by hydration – that is, their final chemical form is calcium sulphate dihydrate ($CaSO_42H_2O$), the same as that of the raw material, though within a crystalline framework.

6.4 Lightweight aggregates

Low density aggregates, such as expanded perlite (produced from siliceous volcanic glass) and cxfoliated vermiculite (produced from mica) form the most important ingredient of modern plasters, which are now almost always premixed. Some advantages of these plasters are as follows:

- Transporting and handling costs of the plaster are reduced.
- The low-density fresh material requires less effort to mix and apply and can be used in thicker coats without sagging.
- The thermal insulation of walls or ceilings is improved and the internal surface temperature increased, thereby improving U values and reducing the risk of surface condensation and pattern staining.
- Fire performance of structures is improved.

6.5 Lime

Lime may be used in quantities up to 25 per cent of gypsum in finish coats of ordinary class B and C plasters. It improves the working properties ('fattiness') of the fresh material and, in class C plasters, counteracts acidity due to accelerators, hence it may help to reduce corrosion of embedded metals. Non-hydraulic limes must be used and should be soaked in water for one day before use.

6.6 Plaster systems

These may be conveniently described under the headings **undercoat** and **finish coat**, although the two components are not completely independent – for example, strong, hard finishing coats also require a reasonably strong undercoat.

Undercoat

The most important factor affecting choice of undercoat is background suction – that is, the tendency of the background to absorb water from the plaster coating. Some suction is desirable, since it removes excess water from the plaster and therefore initiates stiffening. The pores, which are responsible for suction, also increase adhesion of the plaster by a mechanical keying effect.

Table 6.2 *Undercoat plasters: properties and applications*

Background suction	Examples	Lightweight gypsum plaster
Low	Dense concrete; no fines concrete; engineering bricks; plasterboard	Bonding
Medium	Ordinary clay bricks	Browning
High	Some stock bricks; aerated concrete blocks	High suction background browning

Excess suction is a disadvantage because it results in premature stiffening of the plaster, giving too little time for levelling and may lead to poor adhesion. Very low suction is equally a problem, owing to lack of the pores that improve adhesion. Table 6.2 indicates plasters that may be used for backgrounds of various types.

The properties of premixed lightweight undercoat plaster can be varied according to function – for example:

- 'Bonding' plaster contains vermiculite aggregate and produces a relatively dense 'sticky' fresh consistency with good adhesion to smoother, less porous surfaces such as concrete. It can also be used to produce a thicker coat on plasterboard if required.
- 'Browning' plaster contains perlite, producing a consistency that is light and 'fluffy' compared with bonding.
- 'High suction background' plaster contains, in addition, cellulose additives to improve water retention.

The raking of joints in low-suction brickwork or hacking of dense concrete will improve adhesion to backgrounds in the low-suction group. Treatment with PVA bonding agents both reduces water absorption of high-suction backgrounds and improves adhesion of low-suction backgrounds such as high-strength concrete.

Finish coat

This will be selected primarily according to the surface hardness requirement, which will in turn depend on the situation within the building and the function of the building. The most demanding situations are projections in corridors and doorways of public buildings for which the hardest plasters are required. Table 6.3 indicates the gypsum plaster type most appropriate to various situations. Of the plasters listed, the lightweight variety form the largest part of the current market. These do not have the hardness of dense plasters but are nevertheless quite resilient, since impacts are absorbed by localised indentation of the lightweight undercoat.

Table 6.3 *Finish coat plasters: properties and applications*

Function	Plaster type	Comments
Very hard smooth surface	Class D (anhydrite) finish plaster	Strong undercoat needed. Lightweight types not suitable
Hard surface	Class C (anhydrite) finish plaster	Normally on cement/sand or class B/sand backing
Normal purposes/max fire insulation	Lightweight class B plaster	Used on lightweight undercoat

One-coat plasters

Plasters are now available which serve as both undercoat and finishing coat. A typical product consists of a white, high-purity class B gypsum plaster combined with perlite and other additives. The material is best machine mixed. After application to a thickness of about 12 mm, the material is straightened and left for about 1 h to stiffen. The surface is then wetted and trowelled with a wooden float to bring fine material ('fat') to the surface. After a further delay, the plaster can be trowelled to a smooth finish with a metal float. The cost of the material is considerably higher than that of the more common gypsum plasters, but a saving in time can be achieved since 'float' and 'set' functions are obtained in a single coat. The background should be largely of uniform suction (that is, of the same material type) to avoid the difficulty of different areas of the material stiffening at different rates. For example, the stiffening rate would be lower over a concrete lintel compared with lightweight blockwork.

6.7 Plasterboards

These consist of an aerated gypsum core bonded to strong paper liners. Most boards have one ivory-coloured surface for direct decoration and one grey-coloured surface, which has better adhesion properties for plastering. Plasterboards with a foil backing for improved thermal insulation or with a polythene backing for improved vapour resistance are also obtainable.

Plasterboard can be easily fixed as follows:

- To purpose-designed lightweight steel framing using self-tapping screws. This gives a rapid method of production of lightweight internal partitions.
- To timber studding by nails or special screws. Screws are slower to apply but more satisfactory where the timber frame is of low stiffness. (Nailed fixings can sometimes be loosened by vibration as the fixing process continues.)
- To masonry, using dabs of special plaster-based adhesives, provided the background is sufficiently straight.

Table 6.4 *Example dimensions of gypsum plasterboards*

Type	Length (mm)	Width (mm)	Thickness (mm)
Laths	1200	406	9.5 or 12.7
Baseboards	1200	914	9.5
Wallboard	2400	1200	9.5 or 12.7
Plank	1200	600	19.0

Plasterboards offer the advantage of requiring only a very small quantity of water compared to wet applied types; hence they could be used in situations where it is important to keep the relative humidity of the atmosphere to reasonably low levels, where rapid drying is required, or where poor drying conditions prevail. Plasterboard is also available with polystyrene or glass fibre insulation bonded to it, for application to either timber frames or masonry.

A variety of sizes is obtainable (Table 6.4). Laths and baseboards are specifically designed for a plastered finish and, provided joints are staggered, they are less likely than wallboards to result in cracks at joints.

Planks are intended primarily for fire-resistance applications. Various edges are obtainable according to function – for example, tapered edges are suitable for smooth seamless joints on boards to be decorated direct (dry lining), square edges are suitable for plastering, and rounded edges, as in laths, give a good bond to the filler which is used between adjacent pieces.

For plastered finishes, all joints except those in laths must be reinforced with some form of scrim tape to prevent cracking.

The correct procedure is essential in respect of joint treatment to obtain the best finished effect and to avoid cracking.

Where finish coat plastering is to be carried out, this may be in one or two coats and a neat class B 'board finish' plaster is normally used. The thickness applied is quite small – in the region of 5 mm – and it is important that drying is prevented until setting is complete, otherwise a soft, powdery surface will result.

In spite of their paper surfaces, plasterboards provide good fire protection, being designated as class 0 in Building Regulations.

In terms of impact resistance, plasterboards are not as good as traditional in-situ plasters, especially if thin sheets are employed on hollow backgrounds. Thicker sheets, or more frequent fixings, partly overcome such problems. Care is also necessary when making fixings to the material, though reasonably large loads can be carried using special metal or plastic inserts in conjunction with timber battens which spread the load.

Dry lining

In this technique, the ivory face of plasterboard is used as the finished surface, the only 'wet' trade involved being treatment of joints, filling indentations and any small defects.

This offers a further advantage in terms of moisture input into the building. The process cannot, however, be 'rushed' – to obtain a good result, nail or screw indentations must be filled and joints must be taped in the same way as when 'skimming' the whole surface. Filling must be carried in several coats, since the filler material can only be applied in small thicknesses. The final finish is also not as hard as that obtained by application of a thin layer of plaster to the entire surface.

6.8 Plaster for damp applications

It must be emphasised that where damp occurs it is essential to identify and eliminate the cause prior to re-plastering. As indicated in Chapter 4, damp often brings hygroscopic salts to the surface of the wall which chiefly affect decorative finishes such as wallpaper and the plaster surface. These would attract more moisture if left even in relatively dry atmospheres and are therefore best removed. Removal of the old plaster assists in reducing salt levels in the wall. The following could be used after the rectification of damp, since they are able to reduce the effects of any residual damp in the walls:

- A system of dry lining involving fixing (by adhesive) metal channels to the wall and screw fixing special moisture-resistant plasterboard to the channels, followed by usual skimming. This largely isolates the plaster surface from the background and introduces much less water into the plastering process than 'wet' methods.
- Cement/sand/lime plasters form an alternative group of undercoat materials, though these have the disadvantage that finish coat application must be delayed for some days to allow curing and shrinkage of the undercoat. Lightweight premixed versions are available. One possible advantage of cement-based plasters is that they form a barrier to efflorescent salts if these are present in substantial quantities in the background. It is important, when using cement-based undercoats, that their strength does not exceed that of the background, otherwise shrinkage may result in breakdown of the background surface.
- Premixed lightweight renovation plasters (gypsum based) contain additives which assist drying when residual damp is present. They may be used in conjunction with waterproof varieties of PVA-based bonding aids where required.

6.9 Common defects in plastering

These may be associated with background problems, inadequate or incorrect surface preparation, incorrect use of materials or incorrect plastering technique. The main problems are discussed below.

Cracking

Cracks occur when movement of the plaster in excess of its strain capacity occurs. Gypsum plasters are non-shrinking but may still be subject to tensile stress if the background moves. Examples are:

- **Background shrinkage**

  This may result if the background is very wet when the plaster is applied. The plaster then cracks at concave corners or following a definite line of cracks in the background. The only practicable remedy is to fill the cracks and disguise with a suitable decorative finish.

- **Undercoat shrinkage**

  Cement-based undercoats may form numerous hairline cracks if not given sufficient time to shrink before the finish coat is applied. An excess of lime in the finishing coat may have the same effect. These cracks may be filled or simply obscured using a suitable wallpaper. Cement-based plasters tend, in the long term, to form a network of crazing since they have high rigidity and cannot accommodate even quite small movements in the substrate. Such crazing should not noticeably affect the quality of most decorative systems.

- **Plasterboard finishes**

  This type of cracking is widespread, usually following plasterboard joints. The extent of cracking is reduced by use of laths or baseboard, movement then being spread over more joints than in larger sheets. Causes include undersized joists, inadequately fixed or restrained joists, poor nailing technique, omission of scrim tape, inadequate joint filling or simply severe impact or vibration of the structure. Fine cracks can be covered by a textured finish but deeper cracks should be cut out and filled. Where background movement is the cause of the cracking, it will be likely to recur.

- **Structural movement**

  This leads to well-defined cracks which follow a continuous line through the structure, plaster cracking in the same position on each side of solid walls. Cracks tend to be concentrated at weak points such as above doorways or windows. Before cutting out and filling, it is essential to establish and rectify the cause of the movement.

Loss of adhesion

This results when a strong finishing coat is applied to a weak backing coat, especially if the backing is inadequately 'scratched' to form a mechanical key or if it is still 'green'. The problem is uncommon, except with sand/cement backings. Both coats must be replaced unless the problem is caused by a green backing; if this is the case, it should be allowed to harden, loose material removed, and the finish coat reapplied.

The problem will occur with plasterboard if too much lime is added or, in two-coat work, if too much sand is used in the undercoat. The plaster must be stripped off and replaced. If the exposed surface is damaged or uneven, the plasterboard also must be replaced.

Dry out

This occurs if plaster dries before the water becomes chemically bound by setting. It occurs if thin coats are applied to dry backgrounds such as plasterboard, especially in hot, dry weather. During the setting process all gypsum plasters take on a much darker colour and this is replaced by a lighter colour when the hardened material finally dries. If the plaster is allowed to dry before the darker colour stage is reached (usually about two hours) this will signify that the material has 'run out' of the water which it needs for setting. The result will be a soft powdery surface which may be difficult to paint or paper. Defective plaster should be stripped and replaced.

Efflorescence

British Standard 1191 limits the amount of efflorescent salts in gypsum plasters, but salts may be present in the background or in sand (when used) if not clean. Salt deposits may appear on drying of plaster, especially if the background is wet during plastering. Deposition may also result subsequently if a plastered area becomes wet due to a leak. The salts crystallise below the surface of emulsion paints, causing loss of adhesion. They can be removed by brushing, once plaster is dry, and will not recur provided the plaster does not become rewetted.

SELF-ASSESSMENT QUESTIONS

1. Give four advantages of gypsum plasters compared with a lime or cement alternative.

2. Describe the composition, properties and uses of the four main classes of gypsum plaster.

3. Suggest why lightweight premixed gypsum plasters have become very popular in recent years.

4. Explain how the choice of plaster for a given situation will depend on:
   (a) the suction of the background;
   (b) the requirements of the finished surface.

5. Explain the reasons for traditional plastering being carried out in three stages. Give situations in which two- or one-coat work may be acceptable.

6. Describe two possible ways of replastering a solid wall in which dampness (now rectified) was previously present. State the relative merits of each method.

7. (a) Describe the various types of plasterboard that are available, giving applications of each.
   (b) Give reasons for cracking in skim coats to plasterboard and suggest how the risk of cracking can be minimised.

8. Give possible causes of loss of adhesion of the finishing coat on
   (a) solid backgrounds;
   (b) plasterboard backgrounds.
   Give steps that may be taken to avoid or to rectify the problem in each case.

REFERENCE

British Gypsum White Book: *Technical manual of building products*

STANDARD

BS 1191: *Specification for gypsum building plasters*
   Part 1:1973 (1994): *Excluding premixed lightweight plasters*
   Part 2:1973 (1994): *Premixed lightweight plasters*

7 Metals

Chapter summary

- Metal structure and defects.
- Steel composition and properties.
- Steel in construction.
- General applications of metals in construction.
- Corrosion of metals.
- Recycling metals.

7.1 Introduction

In spite of their high cost there will always be a place for metals in the construction industry because

- Metals are generally ductile – they exhibit the phenomenon of being able to undergo substantial plastic flow without damage. In this respect they are unique among stronger materials types.
- They have a wide range of properties according to type.
- The stronger metals have higher stiffness and tensile strength than most non-metals.
- They are non-porous, hence chemical activity tends to be confined to the surface.
- Metals are easily alloyed, further increasing their versatility.
- Strong bonds can be produced easily by soldering, brazing or welding.

There are, however, some problems associated with metals.

- They are expensive in energy terms to produce.
- They deteriorate by chemical surface action in normal atmospheres (corrosion).
- They have high density – even aluminium, with a relative density of 2.7 is denser than dense concrete (relative density about 2.3).

Environmental considerations are given in Table 7.1.

Table 7.1 *Environmental considerations: metals*	
Consideration	Assessment
Raw material availability	Some metals abundant – for example, aluminium and iron. Sources of zinc, chromium, copper and lead are very limited
Extraction	Often considerable scarring of landscape
Energy used in manufacture	Usually energy intensive – especially aluminium
Health/safety hazards	Toxic gases produced by production of some metals. Landfill may contain small amounts of copper, lead and cadmium which are toxic in solution
Recyclability	Metals most frequently used – steel and aluminium are easily recycled at relatively low cost

7.2 Metal structure

It will be recalled that the metallic state is taken up by most elements when the conditions are not suitable for combination with other elements to form ionic or covalent bonds. Hence elements to the left of the bold line of Fig. 2.6 form metals when pure (those to the right, other than inert gases, forming covalent bonds by electron sharing). Many metals to the left of the table are highly reactive on account of having one or two electrons in their outer shell, but most have partially complete subshells, making them less reactive. Many metals are suitable for construction use, aluminium (no. 13) being the commonest and lightest.

Elements 21 to 30 are almost all used in construction. They are known as the transition elements:

- Scandium
- Titanium
- Vanadium
- Chromium
- Manganese
- Iron
- Cobalt
- Nickel
- Copper
- Zinc

Of these, iron has a number of advantages for construction use and fortunately is by far the most abundant of the group (perhaps because it is the most stable element).

One of the big advantages of metals is that the metallic bond is a relatively flexible type of bond. As metallic bonds are not directional, bonding can take place quite readily between most atoms of similar size. Most of the above metallic elements can therefore be bonded with each other, and the process known as 'alloying' can be used to vary the properties of the pure metals.

Mechanical properties of metals related to their structure

It is fortunate that the bond strengths of metals in the above group of transition elements vary considerably (the bond strengths relate to the ease with which each element could obtain a full electron shell and tend generally to reduce as the number of electrons increases). The effect, therefore, is that the tensile strengths broadly reduce with increasing element number in the list and the melting points of these metals also generally get lower. Figure 7.1 shows melting points of the metals. This variation in properties can be very convenient for construction applications. In some applications – for example, sheet roofing – softness and ductility may be a priority, copper and zinc being suitable. In more demanding structural applications iron, with its higher melting point, would be preferred.

Bonding forces, which are non-directional, tend to result in tight packing of atoms and it is known that the tightest arrangement involves hexagons (Fig. 7.2(a)). This is simply the result of the fact that any triplet of touching atoms defines an equilateral triangle with each angle being 60°. Six such triangles define a complete circle of 360°. Hexagons can therefore be found in most metal crystals. Figure 7.2(b) and (c) shows the face-centred cubic crystal, the crystal type exhibited by many metals. At first sight in Fig. 7.2(c) the hexagons are not visible but the planes of atoms defined by the straight lines are in fact hexagonal. (See Experiment 7.1.) These planes are very important because in the face-centred crystal shown, they are relatively widely separated – note the clear space between them in Fig. 7.2(c). This means that shear forces resulting from applied stress would be more likely to cause them to slide (slip) over each other. This is what happens when a metal yields; the most widely separated crystal planes slide over each other. It is important to note that as this happens the bonds continue to reform with

Fig. 7.1 *Melting points of the transition elements.*

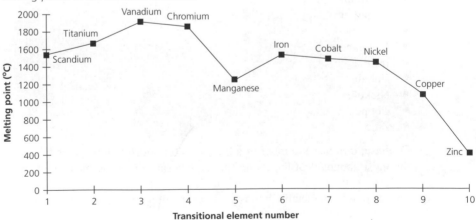

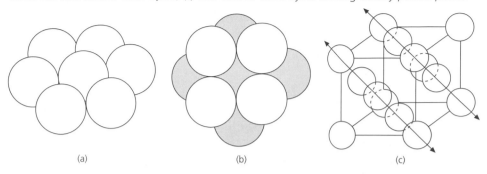

(a)   (b)   (c)

new 'neighbours' since all atoms are the same. Hence, unlike non-metals, yielding does not correspond to 'failure'.

The strength obtainable from metals depends on their ability to resist the effects of shear stresses which are generated by both tensile or compressive stress. This situation can be seen in Fig. 1.6.

It is found that the mechanical performance of metals is reduced greatly from levels which might be predicted from measured bond strengths by imperfections in the metallic crystals.

Grain boundaries

When metals solidify from liquid, the process commences at many different points within the melt more or less simultaneously. These points are known as **nuclei** and, in consequence, a given piece of metal will comprise many thousands of small crystals, originating in these nuclei, which grow until they meet (Fig. 7.3). At this point solidification will be complete. Because each crystal orientation will be different, there will be regions of disorder where they meet. These are known as grain boundaries and are

Fig. 7.3 *The dark lines show grain boundaries and the fine lines, randomly arranged crystal planes.*

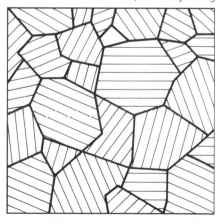

surface defects since the surface of each crystal is affected. In general, faster cooling produces a larger number of smaller crystals. Grain boundaries may be a source of weakness in metals, especially at higher temperatures, on account of the imperfect bonding there.

Dislocations

Dislocations are a most serious form of weakness since they result in yielding of metals at much reduced stresses.

Dislocations result from the rather rapid rates of cooling to which most metals are subjected during normal formation processes. Atoms do not have time to pack perfectly and therefore occasionally planes of atoms come to an end. All bulk metals contain millions of dislocations. When a shear stress is applied, instead of entire planes sliding over each other (which would require thousands of bonds to be breaking and reforming), just one atom per plane swaps neighbours (Fig. 7.4). In the first sketch the extra plane of atoms ends at A (note that all atoms are identical; they are only shaded for clarity and this rectangular representation rather than a hexagonal one is used for simplicity). Under shear stress, atom A bonds to atom B leaving the plane of atoms ending at C as the 'extra' plane (second sketch). The process continues with atom C then

Fig. 7.4 *Schematic illustration of the movement of a dislocation through a metal crystal.*

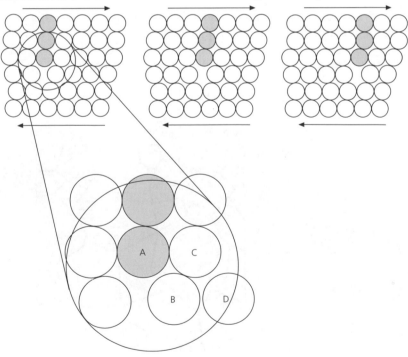

Table 7.2 *Properties of chief building metals (commercially pure)*

In order of increasing ultimate tensile strength

Metal	Relative density	Elastic modulus (kN/mm^2)	Ultimate tensile strength (N/mm^2)
Lead	11.3	16.2	18
Zinc	7.1	90	37
Aluminium	2.7	70.5	45
Copper	8.7	130	210
Iron	7.8	210	540

swapping D for B and so on until the dislocation reaches the end of the crystal. The material has 'yielded' but only by one atom each time on this plane so that yielding has occurred at much reduced stress. Hence, whereas the theoretical yield strength of iron may be 20,000 N/mm^2 or more, the yield stress of the bulk material may be as low as 200 N/mm^2. Dislocations are known as **line defects** since the end of each extra plane of atoms follows a line through the crystal.

It has not yet proved possible to remove dislocations from metals (except in minute crystals which have little relevance to construction), but their effect can be controlled by, for example:

- **Limiting grain size**, which limits the scope for travel of dislocations (dislocations cannot, in general, cross from one grain to another).
- **Work hardening**, in which, by plastic working, dislocations move to the point where they meet obstructions in the metal – for example, grain boundaries.
- **Alloying**, in which crystal planes are interrupted by undulations caused by different sized atoms. Foreign atoms in a crystal are known as **point defects**.

All these are techniques used to improve the properties of metals for construction.

Table 7.2 gives ultimate tensile strengths (that is, breaking strengths) and elastic moduli (stiffnesses) of the most commonly used building metals, together with their relative densities. It will be noted that there is reasonable correlation between strength and stiffness, since both are related to metallic bond strength, values for iron being substantially in excess of those for the other metals. Density is not, however, well correlated to the other properties. The following comments apply to the strength values in the table:

- The tensile strength of metals may depend on the cooling regime – more rapid cooling generally produces a finer grain structure and hence higher strength.
- Yield stresses, which determine the largest load that can be supported in service, are usually substantially less than ultimate stresses.
- Metals such as zinc and lead may, in the long term, 'creep' to failure at much lower stresses than those indicated.
- The presence of impurities or alloying elements generally has a marked effect on strength. Most impurities can increase strength, but they also increase brittleness.

Fig. 7.5 *Magnified photographs of polished and etched sections of steel having various carbon contents: (a) carbon content almost zero; (b) carbon content 0.5 per cent; (c) carbon content 0.83 per cent; (d) carbon content 1.2 per cent.*

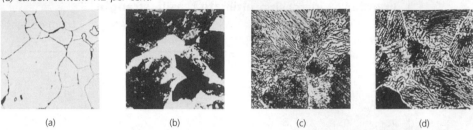

(a) (b) (c) (d)

7.3 Steel

Steel, the most important ferrous metal in building construction, is an alloy of iron and carbon. Its widespread application for structural uses is on account of the following properties:

- Iron is relatively abundant in the earth's crust; and is second only to aluminium.
- Its melting point is sufficiently high to produce high strength, but not so high as to give problems reaching this temperature in bulk furnaces.
- Iron undergoes a change of crystal structure at red heat which assists in many production processes and leads to metallurgical advantages.

Pure iron is crystalline at room temperature, but not in the face-centred cubic form described above, due to covalent influences. The metal instead forms a less ductile body-centred cubic crystal. The material is nevertheless too soft for most applications and, in any case, the product from blast-furnaces – the first stage in steel production – contains substantial quantities of carbon.

The presence of carbon in iron leads, at ordinary temperatures, to the formation of a non-metallic compound – iron carbide (cementite) with the formula Fe_3C, which is intensely hard and brittle, and exists as very fine sheets interwoven with iron sheets to form a compound called pearlite. The advantage of having some pearlite in steel is that dislocation flow is impeded so that plastic flow, or 'yield' occurs much less readily than in pure iron which contains no pearlite. However, the presence of too much pearlite reduces ductility, toughness and weldability. The quantity of pearlite contained in a steel depends on its carbon content. The average carbon content of pearlite is 0.83 per cent (by weight), so that steels having this percentage of carbon would consist of 100 per cent pearlite.

Steels containing less than 0.83 per cent carbon comprise a mixture of pearlite and iron crystals, the amount of pearlite increasing proportionately from 0 to 100 per cent as the carbon content rises from 0 to 0.83 per cent.

Steels may contain up to 1.7 per cent carbon, the excess carbon over 0.83 per cent in steels of carbon content in the range 0.83–1.7 per cent existing in the form of thin cementite layers around pearlite 'crystals'.

Figure 7.5 shows magnified photographs of polished, etched sections of steel having various carbon contents.

- Figure 7.5(a) is amost 100 per cent ferrite – only grain boundaries are visible.
- Figure 7.5(b) is mainly pearlite (layers not resolved in this picture).
- Figure 7.5(c) shows 100 per cent pearlite with layers resolved.
- Figure 7.5(d) shows pearlite with thin (white) cementite layers.

Low-carbon steels

These may be defined as steels containing less than 0.25 per cent carbon and include steels referred to as mild steels and high-yield steels. They form a very large proportion of the steels used in construction, having moderate strength but good toughness, ductility and welding properties on account of the fairly low proportion of pearlite. Examples include weldable structural steels, cold-rolled steel sections for lintels, etc., window frames, reinforcing bars and mesh, expanded metal and wire for wall ties.

In addition to containing some carbon, low-carbon steels contain a number of other elements, some present as a result of the steel-manufacturing process and others included to benefit certain properties of steel.

Some common impurities are:

- *Sulphur and phosphorus.* These form precipitates of iron sulphide (FeS) and iron phosphide (Fe_3P) respectively, which cause embrittlement. Hence contents of each are limited to 0.05 per cent in mild steel.
- *Silicon.* When present in small quantities it dissolves in the iron, increasing strength. Large quantities cause embrittlement, hence silicon content is usually restricted to about 0.2 per cent.

Common alloying elements are:

- *Manganese.* This is useful since it increases the yield strength and hardness of low-carbon steels. It also increases hardenability and may combine with sulphur, reducing brittleness caused by the latter. Mild steel contains approximately 0.5 per cent manganese, while high-yield steel contains about 1.5 per cent.
- *Niobium.* This tends to produce smaller crystals (grains), which results in increased yield stress. Quantities in the region of 0.04 per cent may be present.
- *Others.* In addition to the above, small quantities of molybdenum, nickel, chromium and copper may be included in order to obtain specific properties.

The ladle content of a typical high-yield weldable structural steel might be, for example:

C   = 0.21 per cent
Mn = 1.5 per cent
Cr   = 0.025 per cent
Mo = 0.015 pcr ccnt
Ni   = 0.04 per cent
Cu = 0.04 per cent

Heat treatment of steels

One important reason for the versatility of steel arises from the fact that:

On heating to red heat (about 900 °C), iron changes its crystalline form from body-centred cubic (BCC) to face-centred cubic (FCC).

FCC iron has two important differences from BCC iron:

- It is much more ductile since the crystals contain large numbers of well-defined 'slip planes', as explained above. For this reason, large steel sections are always heated to red heat before rolling or other working because at this temperature they have working characteristics more like lead. The craft of the blacksmith works on the same principle.
- FCC iron dissolves carbon much more readily – up to about 1.7 per cent by weight.

In order to understand how heat treatment can affect the properties of steels, it is necessary to appreciate that the **crystalline form of iron alters each time it is heated to or cooled from temperatures in the region of 900 °C**. Above this temperature the iron crystals ('gamma' iron) can fully dissolve all the carbon present (up to 1.7 per cent), forming a material called **austenite**. Hence, each time steel containing up to 1.7%C is heated, the carbon fully dissolves and each time steel is cooled below about 900 °C, low-temperature ('alpha' iron) crystals form as the carbon is rejected to produce a separate compound – iron carbide or cementite – whose physical form has been described. It will be obvious that the formation of new crystals involves the movement of carbon atoms within the solid matrix of material, and this requires a certain length of time to occur. Hence the properties of the resultant steel can be greatly affected by the rate of cooling. It will also be apparent that steel containing more carbon will be more sensitive to variations in cooling rate since there is more carbon to be rejected (see Experiment 7.2). The main forms of heat treatment are:

Annealing

The steel is heated to about 900 °C so that all carbon dissolves in the high-temperature iron crystals. The steel is then cooled very slowly in the furnace, which results in a coarse crystal structure with relatively thick cementite and iron sheets forming in the pearlite. The result is a soft malleable steel of fairly low yield point (Fig. 7.6(a)).

Normalising

This involves heating the steel as in annealing and allowing the metal to cool in still air rather than in the furnace. The grains so produced are smaller (Fig. 7.6(b)), as are the cementite layers in the pearlite, resulting in increased strength and impact resistance. Hot rolling of structural steel sections, followed by cooling in air, may give properties similar to those obtained by normalising if the 'finishing' temperature is correct; this then avoids the expense of reheating the steel.

Quenching

This involves plunging the red hot metal into cold water, resulting in a very rapid cooling rate. The carbon in this case is unable to form cementite and so a fine, brittle crystalline structure called **martensite**, forms instead (Fig. 7.6(c)). This structure is

Fig. 7.6 *The effect of cooling on low-carbon steel: (a) annealing – very slow cooling in the furnace; (b) normalising – cooling in still air; (c) quenching – rapid cooling in water.*

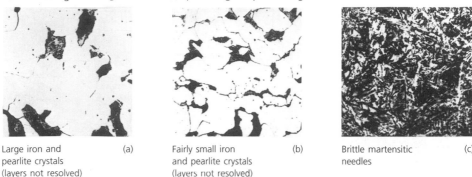

Large iron and   (a)
pearlite crystals
(layers not resolved)

Fairly small iron   (b)
and pearlite crystals
(layers not resolved)

Brittle martensitic   (c)
needles

highly stressed and, in the case of high-carbon steels, cracking may occur in the metal. Quenching is, for this reason, followed by tempering – controlled heating to 200–400 °C, to restore toughness to the steel without losing all the hardness imparted by quenching.

It should be emphasised that the effects of rapid cooling are far greater when steels have higher carbon contents, since high-carbon steels have, on cooling, more carbon to 'reject' from their high-temperature crystal structure. Rapid cooling will also be more effective when applied to small components, since they have lower thermal inertia and therefore cool more quickly in given conditions. Quenching of sections as large as rolled steel joists would be completely impracticable because the material would form a shroud of steam around itself if immersed.

Brittleness in steel

Brittleness implies the inability of the metal to absorb energy by impact (shock loading). Steels which are able to absorb energy in this way would be described as 'tough'. Brittleness depends on:

- **Grain size**
  It is found that, in the case of ferrous metals, a fine grain size reduces brittleness.
- **Carbon content**
  Carbon in the form of cementite increases the resistance of the metal to plastic flow at the point of impact. Hence increased carbon contents tend to increase brittleness, since higher carbon steels contain more cementite.
- **Ambient service temperature**
  Reducing the temperature increases the yield point of steels, therefore making plastic flow more difficult and increasing brittleness.

Eurostandards for structural steel contain a number of toughness categories, the tougher steels being more suitable for service in colder climates or where severe impact is a possibility.

Fig. 7.7 *Elongation at failure of a tensile specimen as an indication of ductility. Elongation around the point of fracture is very high in low-carbon steels.*

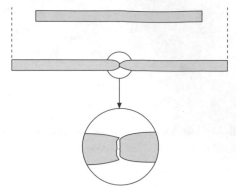

Fig. 7.8 *Stress–strain curves for mild steel, high-tensile steel, copper and lead. The last two reach very high strains before failure.*

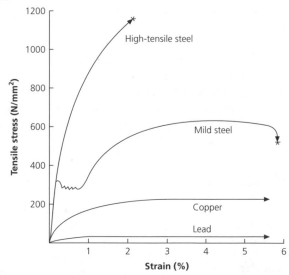

Ductility of steels

This is a measure of the amount of plastic flow that can be absorbed before failure. It is assessed by measuring the elongation at fracture of a standard length of test piece (Fig. 7.7). Ductile materials undergo 'necking' (reduction of cross-sectional area) before failure. Low-carbon steels exhibit highest ductility, especially if annealed, since annealing results in large grains. By carefully controlled cooling, some degree of ductility can also be imparted to high-carbon steels, though these steels are always more brittle than mild steels because they undergo less plastic flow prior to failure. Figure 7.8 shows stress–strain curves for copper and lead for comparison. Note that mild steel shows a very clear 'yield point drop', because carbon atoms in the steel tend to migrate to dislocation sites, occupying positions at the end of extra planes (carbon atoms are very small compared with iron atoms). They 'lock' the dislocations in place and this requires extra stress to cause yield. High-carbon steels do not exhibit a visible yield point.

Weldability in steels

This is the ability of two pieces of steel to be joined by fusion in a localised region at their boundary. The high temperatures involved in welding may result in fortuitous 'heat treatment' of an area of metal in the neighbourhood of the filler metal – the heat-affected zone (HAZ). Where conditions are such that rapid cooling occurs (mainly in large sections), a brittle martensitic structure results, while slower cooling may cause an undesirably coarse grain structure. Such brittleness problems are most severe in steels having a high carbon or alloy content; steels which are designed to be welded would normally have a specified 'carbon equivalent' (CE) content, given by:

$$CE = \%C + \frac{\%Mn}{6} + \frac{\%(Cr + Mo + V)}{5} + \frac{\%(Ni + Cu)}{15}$$

where C = carbon; Mn = manganese; Cr = chromium; Mo = molybdenum; V = vanadium; Ni = nickel; Cu = copper.

The numerical factors applied to the metals other than carbon reflect their decreased tendency to produce brittle martensitic zones.

Few problems should be encountered if the CE is less than 0.25. Higher values of up to, say, 0.5 per cent, may be tolerated if cooling is controlled and precautions are taken to keep down the hydrogen content of the weld metal and the HAZ. Hydrogen can be introduced by moisture in some fluxes and it tends to result in cold cracking unless it is dispersed by heat treatment.

7.4 Steel types used in construction

There are very few buildings in which steel is not used in some location as a structural material. The chief applications are as follows:

- In hot-rolled sections in steel-framed or other buildings.
- As reinforcement or prestressing in concrete-framed buildings.
- As cavity ties and cold-rolled sections for lintels in masonry construction.
- For fixings in timber-framed buildings or timber roofs.

The requirements of steel for use in structural sections, the reinforcement of concrete and the prestressing of concrete are now briefly considered. For the first two applications, the stiffness (E value) of the steel should be as high as possible, since high E values reduce deflections in members. However, E values do not vary greatly according to steel type; in other words, higher strength grades of steel operate at higher strains.

Steel for structural sections (BS EN 10025)

Such sections are invariably hot rolled, taking advantage of the much greater ductility of steel at red heat. The principal requirements of such sections are:

- **Weldability** – welding is an economic and extremely effective means of producing structural joints in steel.
- **Ductility/impact resistance** – steel must be resistant to impact at all service temperatures likely to be encountered, and, if overstressed, should yield rather than fail in a brittle manner.

As far as is consistent with the above requirements, steel should have the highest possible yield strength consistent with economic production.

The former BS 4360 steel grades have now been replaced by the BS EN 10025 equivalents. For example, in a steel designated

S 275 J2 H

'S' refers to structural steel; '275' refers to the minimum yield strength (N/mm^2); 'J2' refers to the impact performance; and 'H' stands for hollow section.

This specification forms the basis of most applications and is roughly equivalent to the previous 'Grade 43D' of BS 4360. The newer description of tensile performance is more appropriate since 275 refers directly to yield performance whereas the earlier '43' was one tenth of ultimate strength, which has much less relevance to structural design.

Higher grades can be obtained by adding alloying elements such as manganese or niobium, or by heat treatment such as normalising. For example, S355 operates at a yield stress about 30 per cent higher than S275. This would permit more slender design or larger spans where required, though higher specifications would normally be supplied to order at increased cost.

Various impact strength options are available according to requirements and these have codes such as JR, JO, J2, K2 etc., in order of increasing performance, which would be combined with the above designations.

All structural steels are of low carbon content, typically 0.2 per cent, to permit welding and produce impact resistance and ductility, and all internal structural steel sections must be adequately protected against fire.

Reinforcing steels (BS 4449; BS 4483)

*When fixed in place in the formwork, reinforcing bars are **passive**; there are no applied stresses.*

Therefore, in order that the steel carries stress in the completed section, loads must be transferred via the concrete. This requires a **bond** between the steel and the concrete. Some of this transfer can be by means of a local bond between the steel and concrete surfaces, though in many cases, hooked ends are provided to prevent steel from slipping under stress. The situation is illustrated in Fig. 7.9. With hooked ends but no local

Fig. 7.9 *Cracking in tensile zone of reinforced concrete beam: (a) if there is no local bond between steel and concrete; (b) if a local bond between steel and concrete exists.*

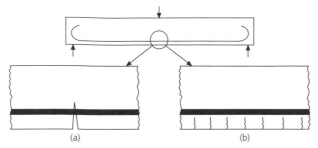

(a)   (b)

bonding the cracking pattern (a) would be obtained in the tension zone, such a large crack tending to admit water and leading to deterioration of steel.

Local bonding would result in cracking pattern (b) which is much more acceptable. Note that since the concrete must move in order to transmit stress to the steel, some cracking in the tension zone is inevitable in reinforced concrete. The safest general approach to reinforced concrete design is to **under-reinforce** the concrete; in other words, the steel yields rather than the concrete failing in shear or compression. If overloading occurs, the steel will then yield but work harden and so stay intact. If the 'weak link' were the concrete, and it then failed in compression, the collapse would be much more dramatic and dangerous.

Hence the requirements for reinforcing steels are:

- *The steel must bond to the concrete.*
- *The steel must be ductile* to permit bending as well as to guard against brittle failure.
- *The steel should be weldable* to facilitate joining (though this is rare on site; lapping is normally sufficient). Fabric reinforcement contains welds at each intersection.
- As far as is consistent with the above, *the steel should have a high yield strength.*

All reinforcing steels are low-carbon steels to obtain ductility and weldability. A 'mild' steel is available to BS 4449 and can be recognised by its smooth round section (Fig. 7.10). The advantage of this steel is that it can form tight bends, but this ductility is obtained at the expense of yield strength, with a characteristic value of 250 N/mm^2. Applications include use in links or where tight bends are required.

High-yield steels are more economic, offering a yield stress of 460 N/mm^2 without a proportionate increase in cost. The minimum bending radius is, however, larger than for grade 250 steel as a result of decreased ductility. Bars are 'deformed' to provide the additional bond required at higher operating stresses. The following types are available according to surface shape:

- **Type 1** These comprise bars which have been cold worked by twisting (Fig. 7.10(b) and (c)).
- **Type 2** (grade 460A) These are hot rolled and are identified by a pattern of two or more parallel straight ribs.
- **Type 2** (grade 460B) These are hot rolled and are identified by a pattern of two or more non-parallel straight ribs (Fig. 7.10(d)).

Fig. 7.10 *Steels for reinforcing and prestressing concrete: (a) plain round bar (BS 4449); (b) cold worked type 1 (BS 4449); (c) cold-worked type 2 (BS 4449); (d) hot-rolled type 2 (BS 4449); (e) 7-wire prestressing strand (BS 5896); (f) alloy steel with prestressing bar with rolled thread and threaded connector.*

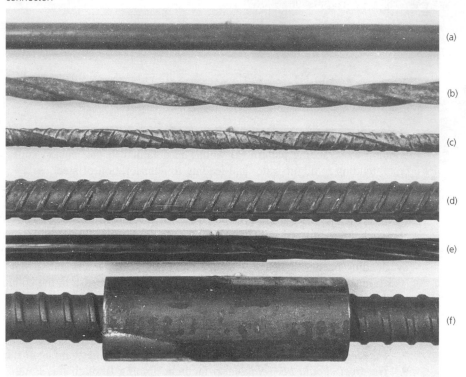

(a)

(b)

(c)

(d)

(e)

(f)

Table 7.3 *Reinforcing steels: properties*

Grade	Yield strength (N/mm²)	Max. strength / Yield strength	Elongation at fracture (%)	Total elongation at maximum force (%)
250	250	1.15	22	–
460A	460	1.05	12	2.5
460B	460	1.08	14	5

Some selected properties of reinforcing steels are shown in Table 7.3. Grades 460A and 460B are given to accommodate requirements in the Eurocode 2 Standard for design of reinforced concrete. The 'A' grade applies mainly to bar sizes up to 16 mm, used for fabric reinforcement. The total elongation at maximum force is a new requirement and will eventually replace the elongation at fracture requirement.

Reinforcing bars are also subjected to fatigue tests, and BS 4449 is soon to be replaced by its European equivalent, BS EN 10080.

For applications such as floor slabs, steel mesh can be used. This comprises steel wires welded at the intersections. The welds provide local bond and anchorage.

Prestressing steels (BS 4486; BS 5896)

Prestressed concrete is so named because stress is applied by the steel to the concrete before the component is loaded.

There are two types: pretensioned and post-tensioned concretes.

Pretensioned concrete
In pretensioned concrete the steel is stretched in a rigid frame prior to placing the concrete (Fig. 7.11). When the concrete has hardened the tension is released causing the steel to 'prestress' the concrete. It does so by bonding to the concrete; there are no anchorages at the ends – indeed, the units can be cut to smaller lengths after hardening if required. A problem of all prestressed concretes is that beams contract due to shrinkage and creep, leading to loss of prestress, since the tension in the steel is only maintained as long as the concrete stays at its original length. For this reason it is desirable that the steel tendons are stretched as much as possible at their working load in order that any contraction of the concrete be small compared with the steel extension. Pretensioned components must be made in a factory; the technique is widely used for manufacturing lintels and floor beams. The requirements for steels for pretensioning of concrete are therefore:

- *Extensibility.* The steel must be stretched as much as possible at its working load.
- *Impact resistance.* The steel must be able to withstand stresses caused by impacts to the structure.

Fig. 7.11 *Principles of prestressed concrete: (a) pretensioned; (b) post-tensioned.*

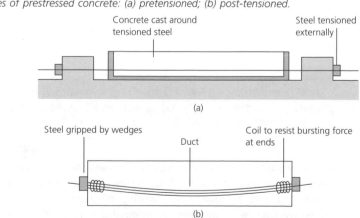

- *The steel should bond to the concrete.*
- As far as is consistent with the above, *the yield strength should be as high as possible*.

Note that welding is not referred to; the material normally used for pretensioning is wire or strand which can be coiled and is therefore available in long lengths. There is also no need to bend prestressing steels. Note also the hexagon shape (Fig. 7.10(e)) which packs wires closely together and forms the basis of many metal crystals, as described above.

The first requirement – extensibility – would be best fulfilled by the use of a low elastic modulus material. The *E* value of steel cannot easily be reduced but because there are no welding or bending requirements, the carbon content can be increased to about 0.8 per cent. This, together with heat treatment, gives a high-tensile steel, which, with unchanged *E* value, also implies a high working strain – over four times that of a mild steel. Typical prestressing wires or strands have a characteristic strength of about 1500 N/mm^2, these steels having a 'springy' feel. In the manufacturing process, the wire, whether used on its own or to form strand, is drawn to increase strength, resulting in a smooth surface. To obtain a bond, wire can be indented. Strand has naturally good bonding properties with the concrete on account of its irregular surface profile.

Post-tensioned concrete

In post-tensioned concrete beams the tendons are enclosed in a sheath or inserted into a duct in the concrete, tensioning being delayed until *after* hardening, since the reaction to the steel load must be taken by the concrete at its ends. The ends are reinforced to withstand the stresses at these positions (Fig. 7.11b). The tendons are not normally bonded to the concrete, hence bonding ability is not required. The extensiblity arguments used for pretensioning steel also apply. Post-tensioning is widely used for in-situ prestressing.

Wire or strand can be used for post-tensioning but for higher load applications high-alloy steel bars can also be used (Fig. 7.10(f)). These cannot be welded but joints can be made since the bars have threads rolled onto them and can be joined by threaded connectors. High-alloy steel prestressing bars of 40 mm diameter have a specified characteristic load of 1300 kN or 130 tonnes – over five times the load capacity of a similarly sized mild-steel bar.

Prestressed units have the advantage that, unlike reinforced concrete, they can be designed to be completely crack-free, compressive stresses being applied to areas of the concrete, which, under load, would be in tension.

Pretensioned units in particular greatly facilitate the construction of openings and of floors. When used in openings, the bricklaying process can continue immediately after placing and in floors an immediate working platform is obtained (the beams would normally be overlaid with an in-situ reinforced concrete slab).

7.5 General applications of metals in construction

Table 7.4 shows typical applications of common metals in construction. Metals for use in any one context will need to be selected on the basis of their performance when measured against the requirements of that situation. A brief summary of possible requirements is given below, with comments on the relative performance of the various metals groups in each case.

Density

There may be situations where additions are to be made to an existing structure, but there are limits to the additional loads that can be applied, and aluminium is the only low-density metal that would be a candidate for such structural modifications. For example, it may be feasible to add a further storey to a building without expensive structural alterations if a lightweight aluminium frame were used.

For sheet applications as in roofing, materials such as zinc and stainless steel would offer advantages over lead as far as weight is concerned (stainless steel can be used in sheets less than 0.5 mm thick).

Durability

Durability requirements of metals should be assessed in the context of

- Cost
- Planned lifetime of the structure
- Operational environment
- Ease of maintenance
- Consequences of deterioration
- Ease of replacement

A most important consideration of metals to be used externally is their operating environment. Although urban atmospheres are in general cleaner than a generation ago, there may still be problems with acid rain, salts in the ground, flue gases and marine environments.

All metals used in construction have at least moderate durability, except in highly exposed or polluted environments, or in situations where electrolytic corrosion is possible. Even unprotected steel corrodes at a slow and predictable rate.

The operating situation should be examined, the possible solutions considered (including factory applied protective systems to metals such as steel) and the most suitable option within cost constraints selected.

Aesthetics

Such considerations can be very important, particularly on more expensive structures. The following levels of performance can be identified:

No loss of surface quality
Metals such as stainless steel (appropriate grade) and bronze can be expected to maintain a clean, high-quality appearance for many years. They do, however, carry a considerable

Table 7.4 *Common metals: properties, recognition and applications*

Metal	Type	Composition	Special properties
Low-carbon steel	Mild steel	Carbon 0.1–0.25% Mn 0.5%	Low cost, good machinability, ductility, weldability, toughness
	High-yield steel	Carbon 0.1–0.25% Mn 1.5%	Greater yield strength than mild steel; ductility not generally as good as mild steel
	Corrosion-resistant steel (Corten)	Up to 0.5% copper. Up to 0.8% chromium	Increased resistance to corrosion – thinner sections possible
	Stainless steel	Contain up to 18% chromium, 10% nickel and other elements such as molybdenum	Highly resistant to corrosion; varying hardness and ductility according to type
Cast iron	Grey	Up to 4% carbon, often high silica content	Low melting point – easily castable; hard, brittle
Aluminium	Pure	99.50–99.99% aluminium	Good corrosion resistance in clean atmospheres (protect from salt, cement mortar, contact with copper); can be hand-formed
	Alloys	Contain manganese silicon, magnesium, copper	Stronger than pure aluminium; can be heat treated
Copper	–	Commercially pure	Ductile but work hardens; noble metal, good corrosion resistance
Brass	Alpha or alpha-beta	Copper alloyed with up to 46% zinc	Lower zinc content brasses are soft and ductile; higher zinc content brasses are more brittle; corrosion resistant
Phosphor-bronze	–	Copper; up to 13% tin; up to 1% phosphorus	High strength and corrosion resistance; can be hardened
Lead	–	Commerically pure	Eminently ductile, does not work harden; very durable (protect from organic acids and cement mortar); creeps; high thermal movement
Zinc	–	Commercially pure	Fairly ductile, moderate durability in clean atmospheres; avoid contact with copper or copper alloys

Recognition	Application Structural	Cladding/roofing	Aesthetic	Manufactured
Easily identified by high density and high stiffness; soon rusts in damp atmosphere; mild steel reinforcing bars normally plain round	Steel sections, reinforcing bars, wire	Plastic-coated steel sheet guttering	Wrought products	Window frames, office furniture, sundries.
As mild steel; high-yield steel reinforcing bars are ribbed; more springy than mild steel	Steel sections, reinforcing bars	–	–	Nuts, bolts, springs
Orange/brown coating, darkening with age to purple/brown	Exposed sections			
Usually 'satin' finish; some types may rust if dirty	Reinforcing bars	Sheet cladding	Balustrades	Tubing, sanitary applications
High density, rough surface texture; brittle fracture revealing grey granular surface	Compression members in older structures	Rainwater goods	Ornamental work – staircases	Boilers, radiators, manhole covers, door furniture
Low density; bright greyish metal surface when new, white deposit when corroded; fairly soft metal	–	Flashings, weatherings, guttering	–	Aluminium foil for insulating boards
As above	Lightweight frames	Roofing, cladding, rainwater goods according to type	–	Door handles
Characteristic 'copper' colour darkening to brown on ageing; green 'patina' formed in dirty atmospheres	–	Sheet roofing, flashings, weatherings	Household utensils	Tubing
Yellow/gold colour, can be polished	–	–	Door furniture	Pipe fittings, hinges, screws, etc.
Dark copper colour	–	Widely used for fixing of claddings	–	Weather proofing strips
High density, dull grey finish, very easily distorted	–	Sheet roofing or cladding, flashings, weatherings	Castings such as rainwater hoppers	Pipes
Higher density than aluminium, fairly soft, light grey finish, becoming powdery with age	–	Roofings, flashings, weatherings	–	Door handles (die cast alloy)

cost penalty. Stainless steel has a bright surface appearance which highlights undulations in a sheet which is not perfectly flat and can cause annoying reflections. For this reason it is often finished with a thin coating of lead (see next category).

Minimal loss of surface quality

Metals in this category include:

- Lead – slight surface patina.
- Copper – green patina.
- Zinc – slight powdering of surface.
- Aluminuim – loss of sheen.
- Weathering steel – brown patina.

The rate at which these processes occur will depend on the operating environment.

Significant corrosion

All remaining metals, or the above in aggressive environments, must come under this heading. Retention of aesthetic qualities will depend upon surface coatings which will require periodic maintenance.

Structural performance and malleability

These two properties must be traded off against each other since they are to a large extent mutually exclusive. In some cases (notably steel) careful use of heat can enable both properties to be obtained. In other cases (for example, copper tube) metals are amenable to work hardening so they could be marketed in hardened form and heated carefully where bending is required. In general, common construction metals (unalloyed) can be formed into a 'league table' of malleabilities as follows:

> Steel
> Copper
> Brass
> Bronze
> Aluminium
> Zinc
> Lead

Loadbearing capacities reduce and malleabilities increase on moving down the table.

Welding/brazing/soldering

There may be situations where watertight and/or structural joints are required between metals, and steel scores very highly in this respect since full strength load transfer can be easily obtained by welding. Where less structural performance is required the metal can also be brazed. In this process the metal is heated (with flux to exclude oxygen) and molten brass applied, wetting the steel surface, then solidifying to give a strong joint.

Most metals can be soldered, but a suitable flux must be used to prevent oxidation. Solders were traditionally lead/tin based, but lead-free versions are now available. Most important uses include capillary fittings for plumbing and lead working for roofing. The joint strength is not high but adequate bond area in each case ensures satisfactory performance. Aluminium cannot readily be soldered because the adherent oxide coating cannot easily be removed. It can, however, be welded using inert gas techniques (normally a factory process).

7.6 Corrosion of metals

One essential feature of the metallic bond is that metal 'ions' are completely surrounded by like ions. As this cannot occur at the surface of a metal, metals consequently have an inherent surface instability. The term 'corrosion' is normally reserved for the deterioration of metals which results from this instability.

Oxidation

In oxygen-containing atmospheres, covalent-type bonds are formed between metals and oxygen where they are in contact at the metal surface. Hence metals in the earth's crust are generally found as 'ores' – combinations of metals and other elements, of which oxygen is the most common. For example, to extract iron from its ore requires a *reduction* process: the ore is heated with coke, which has a strong affinity for oxygen such that the ore is reduced to the metal, while the coke is oxidised to carbon dioxide.

Once exposed to the air, the oxidation process immediately recommences, although in the case of many metals, including iron or steel at ordinary temperatures, this fortunately occurs at a very slow rate. The restriction on the rate of oxidation of steel results from the difficulty of access of the gas to underlying metal atoms. There follows a progressive reduction in oxidation rate such that, after a time, destruction of the metal ceases. Hence, in a dry atmosphere, steel components will remain sensibly free of corrosion for many years.

The main problem with respect to oxidation is that, at high temperatures, the oxide film grows much more quickly. Thicker films tend to crack, thereby exposing fresh metal, which continues to oxidise. Hence substantial oxide coatings (mill scale) can form on steel during hot working. This is recognisable as a thin grey coating on hot-rolled steel beams or reinforcing bars. (In cold-worked reinforcing bars, the mill scale falls off during the deformation process.) Similarly, when metals are being joined by fusion, fluxes must in general be used to prevent oxidation of the hot metal.

Acidic and electrolytic corrosion

This process is quite different from oxidation, although oxygen may assist the corrosion process as, for example, in the rusting of iron or steel. Both acidic and electrolytic corrosion result from a tendency of metals to ionise (dissolve) when placed in water or aqueous solution (Fig. 7.12). This ionisation is the result of interaction between the surface atoms of the metal (which, as has already been explained, are relatively unstable) and ions in the water. The basic process can be represented as:

Fig. 7.12 *Tendency for metals to dissolve when placed in water.*

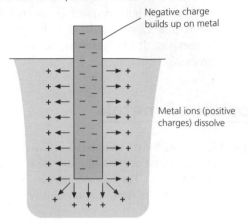

$$M \rightarrow M^+ + e^-$$

surface   positive   electron
metal   metal   (remains
atom   ion   on metal)

The electrons remaining on the metal cause it to become negatively charged, resulting in a negative voltage which increases as further metal dissolves. It eventually reaches a value, called the **electrode potential** of that metal, when the **negative** charge is sufficient to prevent further **positive** ions leaving the metal, since unlike charges attract. Ionisation (corrosion) ceases at this point.

Continued corrosion, resulting in visible loss of metal, will only take place if, by some mechanism, the negative electric charge on the metal is reduced, thereby allowing more positive ions to 'escape'.

In acidic corrosion, the metal is attacked by hydrogen ions in solution.

The hydrogen ions remove the negative charge from the metal and form hydrogen gas:

$$2H^+ + 2e^- \longrightarrow H_2$$

in   from   hydrogen
solution   the   gas
    metal

Metals which, in clean damp conditions, are normally protected by coatings of oxide or other compounds, may nevertheless corrode:

- In the presence of acids such as are obtained in peaty soils (zinc, lead and copper affected).
- When in contact with timbers such as Western red cedar (copper, zinc and lead affected).
- When subject to rain in polluted environments (zinc affected).

Hydrogen is given off **at the point of corrosion** and in severe cases the bubbles may be visible.

Table 7.5 *Standard electrode potentials of pure metals*

Metal	Electrode potential (V)
Magnesium	−2.4
Aluminium	−1.76
Zinc	−0.76
Chromium	−0.65
Iron (ferrous)	−0.44
Nickel	−0.23
Tin	−0.14
Lead	−0.12
Hyrogen (reference)	0.00
Copper (cupric)	+0.34
Silver	+0.80
Gold	+1.4

Fig. 7.13 *Electrolytic corrosion resulting from zinc and copper rods immersed in aqueous solution while in electrical contact.*

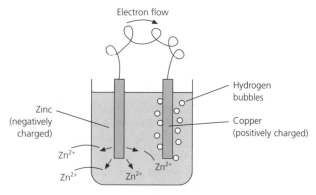

Electrolytic corrosion occurs when two metals which are effectively different are in electrical contact in the presence of water.

This is because the electrode potential varies from one metal type to another. Table 7.5 gives approximate values for common metals.

Taking, for example, zinc and copper in contact (Fig. 7.13), the zinc gives a voltage of −0.76 V, while copper gives a voltage of +0.34 V, each with respect to a 'hydrogen electrode'.

It may be surprising that there should be positive voltages in Table 7.5. This is due to the fact that, to measure a voltage, two electrodes are needed and the other electrode will also have some tendency to dissolve. A porous glass electrode containing H_2 gas is always used as the other ('reference') electrode since it is appropriate for predicting the behaviour of metals in dilute acids. All the figures in Table 7.5 are relative to the H_2

Table 7.6 *Electrolytic corrosion: common causes (two metals)*

Situation	Metal which corrodes	Remedy
Galvanised water cistern or cylinder with copper pipes; traces of copper deposited due to water flow	Zinc corrodes at the point of contact with the copper particles. Film is destroyed and steel corrodes similarly	Use a sacrificial anode in system. Otherwise use a plastic cistern/copper cylinder
Brass plumbing fittings in certain types of water	Zinc (dezincification)	Use low zinc content brass or gunmetal fittings
Copper ballcock soldered to brass arm	Corrosion of solder occurs in damp atmosphere resulting in fracture of joint	Use plastic ball on ballcock
Copper flashing secured by steel nails	Steel corrodes rapidly	Use copper tacks for securing copper sheet
Iron or steel railings set in stone plinth using lead	Steel corrodes near base	Ensure that the steel is effectively protected by paint
Steel radiators with copper pipes	Steel radiators corrode	Corrosion can be reduced by means of inhibitors

electrode which may be regarded as a 'metal', also having some tendency to dissolve in water. Metals above hydrogen in the table have a greater tendency than hydrogen to dissolve, while those below hydrogen have a smaller tendency than hydrogen to dissolve – indeed, acidic corrosion can be regarded as 'electrolytic' corrosion with hydrogen as the other 'electrode'.

In the case of zinc and copper in contact in water, electrons will flow from the zinc to the copper at their point of contact (conventional current in the reverse direction) such that the negative voltage on the zinc is partially cancelled and the zinc continues to corrode. The zinc is called the **anode** and the copper, which cannot corrode, is called the **cathode**. Hence, electrolytic corrosion will occur between any two metals in contact if moisture is present, and the metal that is higher in Table 7.5 will corrode. The corrosion will be more serious:

- If the anode area is small and the cathode area is large (e.g. galvanised nails in copper sheet). (The anode can, in a sense, be regarded as a 'meal' and the cathode the 'appetite' since the cathode effectively is responsible for consuming the anode.)
- If salts are present to assist ion movement in solution.
- If the temperature is raised since all chemical activity is accelerated when thermal energy is increased – for example, in central heating systems. Conversely, corrosion rates reduce when temperature is lowered and if water freezes corrosion is negligible because ion movement is no longer possible.

Table 7.6 gives some common situations in which electrolytic corrosion can occur.

Table 7.7 Electrolytic corrosion: within a single piece of metal

Cause	Anode	Examples	Remedy
Grain structure of metals	Grain boundary	Any steel component subject to dampness	Protect steel from dampness
Variations in concentration of electrolyte	Low concentration areas	All types of soil	Protective coating or cathodic protection
Differential aeration of a metal surface	Oxygen remote area	Improperly protected underground steel pipes	Protective coating or cathodic protection
Dirt or scale	Dirty area (oxygen remote)	Exposure of some types of stainless steel to atmospheric dirt	Use more resistant grade or keep surface clean
Stressed areas	Most heavily stressed region	Steel rivets	Protect from dampness

It is very important to appreciate that corrosion can occur within a piece of a single metal type, since its structure and also possibly environmental variations may result in effectively different electrode potentials at different parts of the surface. Table 7.7 indicates a number of ways in which this may occur. Steel readily corrodes in damp environments by mechanisms such as these.

Oxygen assists in the corrosion of steel by combining with electrons from cathodes to form the hydroxyl ions necessary to produce rust (ferric hydroxide):

$$2H_2O + O_2 + \underset{\substack{\text{from}\\ \text{cathode}}}{4e^-} \longrightarrow \underset{\substack{\text{hydroxyl}\\ \text{ions}}}{4(OH)^-}$$

The corrosion of steel is also greatly increased if salts such as sodium chloride are present; a very small concentration increases the electrical conductivity of the electrolyte by many times, facilitating the flow of ions in solution. Salts may also be directly involved in the corrosion process. Electrolytic corrosion is by far the most serious form of corrosion of metals in general building applications.

Prevention of corrosion

It will be apparent that, since both acidic and electrolytic corrosion involve ionisation and moisture is necessary for ionisation, each can be prevented by keeping the metal dry. Where dampness cannot be avoided, impermeable coatings will greatly reduce the likelihood of corrosion. However, no coating is absolutely waterproof. The resistance to

Fig. 7.14 *Protection provided by galvanising at breaks in the zinc coating.*

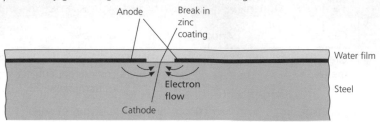

moisture penetration of a paint film increases with its thickness; therefore, situations of exposure to longer term moisture require thicker paint films.

> *Coatings for metals which have a reputation for good corrosion proofing properties are largely those that are amenable to application in thick coats.*

Paint films must be the subject of regular maintenance checks, with more aggressive environments calling for thicker coatings or more frequent checks. Impregnated protective tapes might be preferred for high-risk situations since they can be used to provide a thick film in a single operation, there being no need for a drying period. They are also more resistant to abrasion damage.

Many metals, such as aluminium, zinc, lead and copper, are protected by thin oxide or corrosion films, hence they may not need to be painted in clean environments. Corrosion may, however, be promoted by rain in polluted atmospheres or by contact with acid soils.

There are other means by which corrosion can be prevented or reduced – for example, galvanising of steel, in which a thin zinc coating is applied. The steel is protected for a time, even if exposed by scratches in the zinc. The zinc corrodes in the region of scratches when damp is present, supplying electrons to the steel and hence maintaining it at a sufficient negative voltage to prevent corrosion. On continued corrosion of the zinc, the exposed area of steel increases until it is too large for the zinc to give full protection (Fig. 7.14). Thick coatings, as obtained by hot dip galvanising (approximately 100 μm), are necessary if effective long-term protection is to be achieved without painting. Processes such as electroplating, which is often applied to articles such as nuts and bolts, give much lower thicknesses and consequently lower protection. As with other corrosion processes, the life of galvanised articles depends very much on the presence of corrosive agencies, such as pollution. In clean atmospheres, galvanised steel may last 50 years or more, but in polluted or salty environments, coatings may only last a few years. Zinc coatings are best used in conjunction with another protective system such as paint if best results are to be obtained.

A further possibility is to subject the exposed metal to a negative charge, using a direct current supply (Fig. 7.15). The effect is similar to galvanising except that the negative charge is supplied direct rather than by corrosion of a zinc coating. This method is used for the protection of underground steel pipelines, though it is designed to prevent corrosion occurring in defects in a protective coating rather than to obviate

Fig. 7.15 *Protection of an underground steel pipe by the impressed current method.*

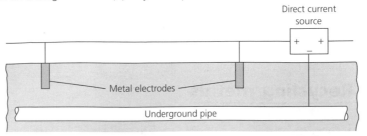

the need for a protective film. Checking the condition of such films in a buried pipe would clearly be very difficult.

The technique can be also used for protecting steel reinforcement in carbonated concrete. In such cases, it is very important to employ closely spaced electrodes in order that all embedded metal be sufficiently negatively charged relative to its **immediate** environment. In the case of reinforcing steel, the 'ideal' is an electrically conductive paint applied to the concrete surface and connected to the positive side of the direct current electrical supply. The reinforcing bars would be connected to the negative terminal. Such systems have been used to protect structures such as bridges, which are often highly exposed. Where signs of distress are observed, cathodic protection can prolong the life of such structures which, in urban environments for example, could cause great expense and disruption if demolished and replaced.

Stainless steels

These are produced by alloying steel with chromium which, if present in sufficient quantities (at least 10 per cent), forms a coherent, self-healing oxide film on the surface of the metal.

Best resistance to corrosion is obtained with a single-phase material, that is, with carbon fully dissolved. (It will be recalled that ordinary steel at room temperature comprises a mixture of ferrite and pearlite.) This means preserving the high-temperature form of iron (see page 192) which can be achieved using nickel – a metal which, like high-temperature iron, has a face-centred cubic crystal structure. Another name for high-temperature iron is austenite; hence, stainless steels containing nickel are referred to as 'austenitic' stainless steels, and a typical example contains:

18 per cent chromium and 11 per cent nickel

Welding of this steel may cause problems due to carbon migration at high temperatures, but steels can be stabilised by the addition of niobium or titanium. The addition of molybdenum can impart even higher corrosion resistance, as might be required in aggressive environments or where cleaning is not possible; for example:

17 per cent chromium, 11 per cent nickel and 3 per cent molybdenum.

By using a **low-carbon content steel in conjunction with about 13 per cent chromium**, a lower cost material with good formability – known as ferritic stainless steel – can be achieved. It is used for items such as balustrades and steel sinks and can be distinguished from austenitic stainless steels because it is magnetic, whereas the latter are non-magnetic (not attracted by a magnet).

Use of a **higher carbon content with 13 per cent chromium** gives a harder (martensitic) stainless steel. Such steels can be used for cutlery or tools.

7.7 Recycling metals

There are compelling reasons for recycling metals due to their very high energy content and the fact that some production processes can lead to waste heat emissions and/or emission of toxic gases or other pollutants.

For example, cadmium, while not used in construction, is highly toxic and may be emitted during manufacture of copper, zinc and lead from their ores. Many manufacturers are now aware of environmental considerations during production, but the position regarding imported metals is much less clear.

Quite separate to the above, metals were one of the earliest materials to become the subject of organised recycling programmes since:

- Identification and separation of most metal types from each other and from other waste is relatively easy. Steel is particularly easy to separate from non-ferrous metals because it is magnetic.
- The energy involved in reheating metals to their melting point is much lower than the chemical/electrical energy used to form them.

Those applications in building in which metals are used in relatively pure forms and those in which they are used discreetly are, from an environmental point of view, to be preferred. Mild steel is easily recycled whereas solders, brass and aluminium alloys containing significant amounts of alloying elements add to the complexity of the recycling process. Similarly, chrome or zinc plating, though they may increase the lifetime of components, cannot be considered to be environmentally 'friendly' processes since there is little chance of recovering the minor constituents on recycling.

Some specific examples are given below.

Cast iron

Uses of this material are now restricted on account of its limited structural performance, though it has the advantage in many applications of being maintenance free. It also has recycling advantages in that the properties of the material are not highly sensitive to small compositional changes and it has a relatively low melting point.

Lead

Use of lead is now much more restricted on account of the health hazard associated with lead in the atmosphere and when dissolved in water. Nevertheless, the durability of lead offers advantages where safely used. Its intrinsic value and ease of recycling has given rise to a high percentage of the material being recycled.

Aluminium

The recycling cost of aluminium is only about 5 per cent of the original manufacturing cost and is a major attraction of the material.

EXPERIMENTS

Experiment 7.1 Modelling demonstration of the face-centred cubic crystal

This experiment demonstrates the packing properties of spheres and, in particular, the link between hexagonal packing patterns and the face-centred cubic crystal – the basis of many common metals.

Apparatus
Supplies of polystyrene balls; sizes of 20 or 25 mm will suffice. Adhesive – 'Superglue' or (slower setting), PVA emulsion adhesive. If superglue is used **precautions must be taken to avoid bonding of skin**. The simplest way to achieve this is to wear rubber gloves. Emulsion paint of a distinctive colour (or oil paint, though these can be slow to dry).

Procedure
Paint about 20 balls with the distinctively coloured paint and allow to dry. The procedure will be to produce the models shown in Fig. 7.16. **Only minute amounts of adhesive are necessary**. Superglues give an almost instant bond, but care must be taken to avoid bonding by accident. Emulsion glues must be given some time to set.

(1) Figure 7.16(a)
Produce a hexagon (six balls around a central ball) with two coloured balls opposite to each other. Produce two triangles each having three balls and with one coloured ball in each.

Fig. 7.16 *Various representations of the face centred cubic metal crystal.*

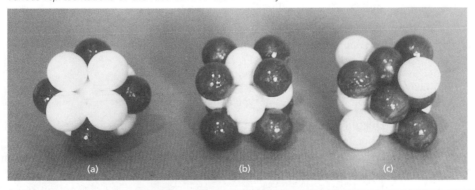

(a)   (b)   (c)

Once dry, place the first triangle on a flat surface with the coloured ball towards you. Place the hexagon centrally on top of it with two coloured balls parallel to the edge of the table. The central hexagon ball should fit in the middle of the triangle. Place the other triangle on top of the hexagon with the coloured ball away from you. The four coloured balls define a cube with the central ball of the hexagon at the centre of the face.

(2) Figure 7.16(b)
This is the same crystal structure but with the emphasis on the cubic nature of the crystal. Glue four coloured balls equally spaced around a white ball on a flat surface to form a cube. Repeat to form a second set. Place four white balls in the four gaps between the coloured balls of one set. The white balls should touch. Finally place the second group centrally on top of the four white balls so that they fit into the gaps. The face-centred cubic structure will now be clear – one white ball in the middle of each face. The hexagon planes in this model may not be clear but their presence will be demonstrated in the final model.

(3) Figure 7.16(c)
Produce a group of five balls as in (2) but with only one of the outer balls coloured. Add four more in the gaps as in (2) but use two coloured balls, one on each side of the coloured ball in the lower layer. Produce a further group of five balls with a line of three coloured. Place this on top of the group of four with the coloured balls diagonally above the coloured ones in the lower layer. The six coloured balls now form part of one of the hexagonal planes. Note how they touch each other. The face-centred cubic crystal actually contains **four** sets of non-parallel hexagonal planes. See if you can find others.

Experiment 7.2   The effects of heat treatment on steel of differing carbon content

Apparatus
Samples of steel wire of low- and high-carbon content (diameters should be the same, values of 2 and 3 mm are suitable); Bunsen burners or blow torches; large beaker; two pairs of pliers; tensometer (if available); small hacksaw and supply of new blades; wire cutters; protective gloves.

Procedure
Note: precautions must be taken against injury resulting from the use of cutting tools and heat.

(1) Untreated properties
Cut suitable lengths of each type of wire and compare:

(a)   Ductilities by attempting to bend and rebend wires until failure; the number of rebends to cause failure may be noted.
(b)   Hardness by cutting through using a hacksaw with a new blade; the number of strokes to cut through may be noted.

(c)   Tensile properties. Ideally a tensometer should be used. A low-carbon steel will show a definite yield point if a load extension graph is plotted (Fig. 7.8). High-carbon steel wire will show no such yield point and will sustain a much higher load for a given diameter before failure (Fig. 7.8). If tensile testing apparatus is not available, the difficulty of producing a permanent bend in the two types of wire, using pliers, can be compared.

(2) Properties of the quenched material
Position a large beaker of cold water near two or more Bunsen burners set up so that they can be used to heat a 150 mm length of wire. Hold the low-carbon steel wire in the pliers and heat the end until it glows red hot. 'Soak' it in the heat for about one minute then plunge the sample immediately into cold water. Repeat for the high-carbon steel wire. Test both samples as described in (1) and compare the properties.

(3) Properties of the annealed material
Repeat (2), but instead of plunging the red hot samples into cold water, withdraw them as slowly as possible from the Bunsen flame by lifting them vertically upwards. Repeat the tests of (1) on these samples.

(4) The effect of quenching followed by annealing
To demonstrate that quenching does not chemically alter the steels (though cracking may result in objects of large section) quench samples as in (2) and then anneal as in (3) before testing. Compare results with those for the samples previously subjected to operation (3) only.

Conclusion
Comment on the effects of each process on the properties of the low- and high-carbon steel. Note that most wires of this diameter are manufactured by drawing through a die. This produces **work hardening** so that annealing may result in reduced tensile performance compared to the untreated wire.

1. (a) Give characteristic properties of:
       (i) iron;
       (ii) cementite.
   (b) Describe the physical form of pearlite.
   (c) State how the proportions of pearlite and cementite in steel depend on its carbon content.

2. Describe the essential properties of low-carbon (mild) steels.
   Explain why a high-carbon steel would not be a suitable material for reinforcement of concrete.

3. Describe the effects of the following elements on the properties of steel, giving typical percentages present in mild steel:

(a) sulphur;
(b) silicon;
(c) manganese;
(d) niobium.

4. (a) State, with reasons, which type of steel is more susceptible to heat treatment – low-carbon or high-carbon steel.
   (b) Describe briefly the following heat-treatment processes:
      (i) annealing;
      (ii) normalising;
      (iii) quenching.

5. (a) Explain how the cooling rate of a welded steel joint affects the performance of the weld.
   (b) Indicate the importance of the 'carbon equivalent' of a steel in relation to welding.

6. A Grade 50C steel gives a ladle analysis as follows:

Carbon	0.21%
Manganese	1.5%
Chromium	0.025%
Molybdenum	0.015%
Nickel	0.04%
Copper	0.04%

Calculate the carbon equivalent of the steel and hence comment on its weldability.

7. State the three mechanisms by which metals may corrode. Indicate with examples which of these is responsible for most damage to metals used in buildings.

8. Give three examples of situations in buildings where electrolytic corrosion may result from the use of different metals in contact.

9. Give three mechanisms by which electrolytic corrosion can occur within a single piece of metal such as steel in a damp situation.

10. Explain the function of traditional oil paints in the protection of metals from corrosion. State precautions which should be taken to ensure that paint fulfils this function effectively in practice.

11. State characteristic properties or features of the following metals which would enable them to be identified:
    (a) high-tensile steel;
    (b) copper;
    (c) brass;
    (d) phosphor bronze;
    (e) zinc.

Give applications of each in the building industry.

12. Comment on the following in relation to recycling of metals:
    (a) attractions of recycling;
    (b) ease of recycling various metals;
    (c) recycling of alloys.

STANDARDS

BS 4449: 1997: *Carbon steel bars for the reinforcement of concrete*
BS 4483: 1985: *Steel fabric for the reinforcement of concrete*
BS 4486: 1980: *Hot rolled and hot rolled and processed high tensile alloy steel bars for prestressing of concrete*
BS 5896: 1980: *High tensile steel wire and strand for prestressing of concrete*
BS EN 10025: 1993: *Hot rolled products of non-alloy structural steels. Technical delivery conditions*
DD EN 10080: 1996: *Steel for the reinforcement of concrete, weldable ribbed reinforcing steel B500. Technical delivery condition for bars, coils and welded fabric*

8 Timber

Chapter summary

- Structure of wood.
- Effects of moisture in wood.
- Conversion of timber.
- Durability of timber.
- Timber preservatives.
- Fire resistance.
- Timber for structural uses.
- Timber products.
- Environmental considerations.

8.1 Introduction

Although one of the earliest materials to be used in building, timber is by no means outdated and continues to play a major part in general building, particularly in domestic dwellings and furniture. More recently, use in larger public, commercial and industrial buildings in the form of timber frames, trussed rafters and glued laminated sections has increased.

Timber and timber products exhibit a number of attractive properties:

- They are natural products, hence they are a renewable resource. With properly managed forestry policies trees can be effective in reducing the carbon dioxide levels in the atmosphere, thereby reducing global warming.
- Manufacture of timber products requires much less energy than alternative materials. Energy required is confined to felling, conversion, machining and transport.
- The material does not pose any disposal problem after use. It rots naturally, reverting to carbon dioxide enabling the growth cycle to be repeated.
- Timber has a high strength to weight ratio and for this reason is particularly useful for roofing where superimposed loads are limited.
- Timber is of aesthetically pleasing appearance.
- Timber has a low thermal conductivity, hence it has a warm 'feel'.
- The material is easily worked with hand tools and joined with nails or screws.

Table 8.1 *Environmental considerations: timber*

Consideration	Assessment
Raw material availability	Very widespread (but not always well managed)
Extraction	May have environmental implications
Energy used in manufacture	Very low
Health/safety hazards	Few direct problems. Wood dust may be a risk. Hazards linked to preservatives and adhesives
Recyclability	Various options available

- Strong joints can be formed with adhesives. Modern adhesives penetrate the wood and many are both strong and waterproof once set.
- Structural timber sections have good fire resistance.

Problems associated with the use of timber are:

- The material may be very variable due to the effects of 'imperfections' such as knots, grain and defects such as 'shakes'.
- Timber is subject to movement and possible distortion when its moisture content changes.
- Timber is subject to fungal and insect attack.
- Untreated timber burns readily in fire and therefore constitutes a fire hazard, contributing to the fuel load of the building and spreading the fire by flame.
- Timber is subject to creep under load, especially if its moisture content is subject to change.

Nevertheless, with careful selection and appropriate use, timber and timber products constitute a very valuable range of building materials. Environmental considerations are given in Table 8.1.

8.2 Structure of wood

Wood is a naturally occurring fibrous composite, the fibres consisting chiefly of crystalline cellulose $(C_6H_{10}O_5)_n$ chains of high molecular weight bonded by non-crystalline hemicellulose and lignin adhesives to form rigid, though largely hollow, cells. It is worth noting from the above formula that carbon forms a major part of the wood fibres, timber being a very important means of reducing atmospheric carbon dioxide. It might be argued that the substance of timber originates mainly from sap drawn from the ground, but it should be appreciated that sap comprises **mineral** solutions (contains no carbon) which largely catalyse the growth. The carbon component is taken from the atmosphere. Hence, as long as the timber is preserved it helps to reduce atmospheric carbon dioxide. (Remember that if the timber is ultimately burned or allowed to rot, it reverts to atmospheric carbon dioxide and the benefit is lost, though the disposal of timber in any form poses much less of a problem than the disposal of most other materials.)

The cells differ considerably from one another according to the classification of timber (whether softwood or hardwood), the particular species and their various functions within each tree.

Softwoods

Figure 8.1 shows a cut portion of a 5-year-old softwood log. Three directions are defined:

- **Longitudinal** – the direction of growth of the tree.
- **Radial** – running outwards towards the bark.
- **Tangential** – running around the perimeter of the tree.

It is evident that cells run chiefly in two directions – longitudinally and radially. Most cells run in the longitudinal direction in order to support the tree and to provide food conduction. Further food conduction and storage are provided by rays which run in the radial direction. Rays also increase the strength of the timber in this direction.

In softwoods the function of both food conduction and support are mainly fulfilled by a type of cell known as a tracheid. Additionally, in the rays there are hollow food storage cells known as parenchyma. Figure 8.2(a) shows a simplified radial section of the softwood cell structure. Adjacent cells are joined in all directions by pits which are responsible for the transmission of food between them. The rays in softwood are usually small and not visible to the naked eye. Figure 8.3 shows a scanning electron microscope photograph of cut transverse, radial and tangential surfaces of a sample of European redwood (*Pinus sylvestris*). Note the resin ducts which are found in some softwoods and the variation in thickness of the tracheid cell walls between spring (rapid growth) and summer (slow growth).

Fig. 8.1 *Radial, tangential and transverse sections, through a 5-year-old softwood log.*

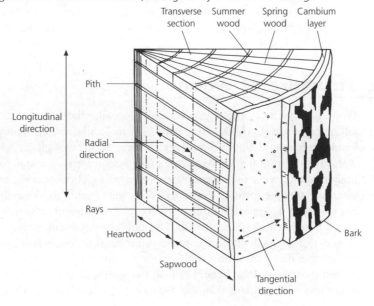

Fig. 8.2 *Simplified radial sections: (a) softwood; (b) hardwood.*

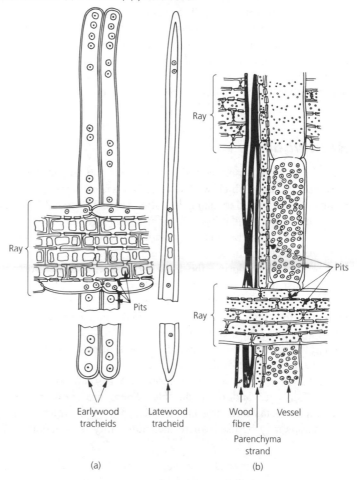

Ray

Ray

Ray

Pits

Pits

Earlywood
tracheids

Latewood
tracheid

Wood
fibre

Parenchyma
strand

Vessel

(a)

(b)

Fig. 8.3 *Scanning electron microscope photograph of European redwood (Pinus sylvestris), a common softwood. The resin ducts form mainly in summerwood. (×10 magnification.)*

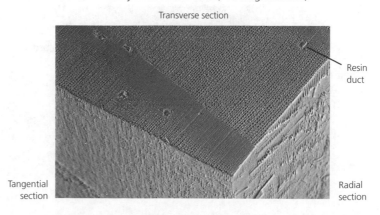

Transverse section

Resin
duct

Tangential
section

Radial
section

Softwood timbers are coniferous (cone bearing) and have needle or fern-like foliage which is retained in winter. They grow relatively quickly, the resultant wood generally being soft, of low density and easily worked. Softwoods are substantially more economical than hardwoods though they are in general less durable. They are now the most widely used type of wood for general structural purposes. Commonly used species in the UK are Scots pine (redwood), European spruce (whitewood), Douglas fir and cedar. Note that the latter wood contains natural resins which make it unattractive to fungi and insects. It has therefore very good durability.

Hardwoods

Hardwoods have a more complex cell structure than softwoods, structural support being provided by long, thick-walled cells known as **fibres**, while food conduction is by means of thin-walled tubular cells called **vessels** (Fig. 8.2(b)). Parenchyma (food storage cells) are again present both as longitudinal strands and in rays, the latter often being visually much more pronounced than in softwoods. Cells are again interconnected by pits. Hardwoods grown in temperate climates such as the UK shed their leaves in winter. Sap requirements during the growing season are therefore considerable, since new leaves must be formed. Hence vessels in springwood are often quite large and easily recognised in cross-sections of the tree. Hardwoods can be subdivided into **ring-porous** varieties, in which the vessels are mainly concentrated in springwood, and **diffuse-porous** woods, in which there is a more even distribution of vessels. Figure 8.4 shows a scanning electron microscope photograph of oak – a typical ring porous wood. Notice also the prominence of the rays in the wood.

Hardwoods generally grow more slowly than softwoods and are usually harder, denser, stronger and more durable. There are, however, exceptions such as balsa and poplar – fast-growing hardwoods with poor structural properties and durability. Hardwoods may be used where their durability is beneficial – for example, in window

Fig. 8.4 *Scanning electron microscope photograph of oak (Quercus robus), a ring porous hardwood. There are broad rays between each group of about 4 vessels in springwood. Most of the radial section in this picture is occupied by a ray with a clear horizontal grain. (×10 magnification.)*

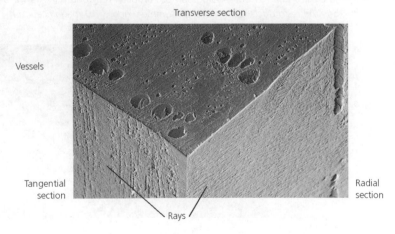

Transverse section

Vessels

Tangential section

Radial section

Rays

---

and door sills; or in high-quality joinery. They are also extensively used in producing veneers for doors and furniture.

Heartwood and sapwood

The food conduction and storage functions of trees are fulfilled by the outer, more newly grown layers of cells, referred to as **sapwood**, the inner layers being referred to as **heartwood**. The latter no longer stores food but is still important with regard to support of the tree, the cells tending to become acidic in order to assist preservation of the tree. Heartwood is normally darker in colour than sapwood and, on account of its acidity, tends to be more resistant to fungi and insects in service. In young trees most of the wood is sapwood, while in old trees as much as 80 per cent of the wood is heartwood. Bearing in mind the difference in durability of the two types, it will be evident that older trees tend to give timber of a better quality. Traditionally sapwood was often rejected, but since this would be uneconomic when, as is now common, young trees are employed, other means of ensuring durability may be advisable such as the use of preservatives. (See Experiment 8.1 for a detailed treatment of the identification of wood species.)

8.3 The effects of moisture in timber

The moisture content of timber is defined as:

$$moisture\ content = \frac{mass\ of\ water\ present\ in\ a\ sample}{mass\ of\ that\ sample\ when\ oven\ dry} \times 100$$

It will be noticed that, if the mass of water contained exceeds the mass of dry timber, a moisture content of over 100 per cent is obtained and this is not unusual in newly felled (green) timber. The sapwood tends to contain more moisture at the time of felling, the average moisture content being about 130 per cent, though values of over 200 per cent are possible. In heartwood the moisture content is more likely to be around 50 per cent. The moisture content of green timber varies considerably with species. Denser timbers, which include most hardwoods, have smaller cavities and larger cell walls, hence they can store less water.

The moisture content of timber in service is greatly influenced by the fact that wood is hygroscopic, that is, it tends to attract water from a damp atmosphere and give up water to a dry atmosphere. In consequence timber will adopt an equilibrium moisture content in a given environment, depending on the relative humidity of that environment. Since increase of temperature tends to reduce relative humidity, it also results in reduced equilibrium moisture contents. Figure 8.5 shows typical values for softwoods in buildings. At moisture contents in excess of 20 per cent, timber becomes susceptible to fungal attack, hence the importance of providing a dry environment will be appreciated.

Recent surveys have shown that timber in service often has a lower moisture content than was formerly assumed in design. Major factors in this are central heating and air

Fig. 8.5 *Moisture contents of timber in various environments (per cent).*

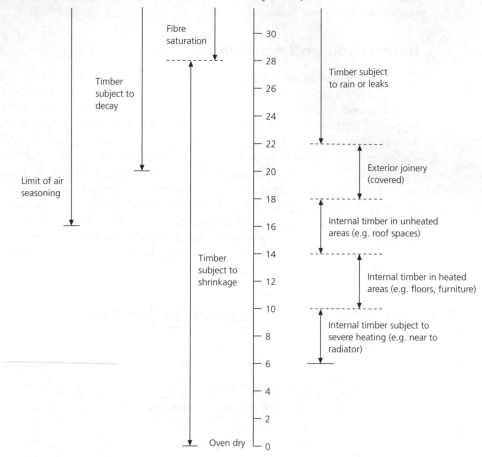

conditioning. In winter cold air is admitted by means of ventilation and this can only contain limited quantities of water. On heating, the relative humidity therefore drops to quite a low value, the moisture content of timber falling in consequence. In very warm buildings such as hospitals the moisture content may drop to as low as 8 per cent, the lowest values being reached in cold dry spells. In normally ventilated buildings which are centrally heated, moisture contents tend to fall towards the end of the winter period. In summer, air-conditioning systems may similarly reduce moisture contents of timber since they condense water out during the cooling cycle.

The only completely accurate way of measuring the moisture content of timber is to oven dry it at $103 \pm 2\ °C$ until the mass is sufficiently constant to give moisture contents within 0.1% of each other. However, this method requires a disposable sample of the wood (not at the end of the piece where the moisture content may not be representative of the piece). It is also rather slow and this level of accuracy is not often required.

The moisture content of wood can be much more easily estimated by means of portable meters based upon the relationship between electrical resistance and moisture

Fig. 8.6 *The effect of drying various sections cut from a log.*

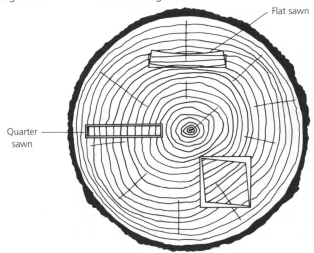

Flat sawn

Quarter sawn

content. A two-pin electrode is pushed into the wood. At lower moisture contents electrical resistance is greatly increased. When using these meters, note the following:

- Calibration depends on species type; readings may vary typically by 2 per cent for a given 'true' moisture content.
- The reading depends on temperature. For example, **reducing** the temperature by 10 °C would **reduce** the apparent reading by about 1.3 per cent moisture content.
- The reading obtained will be an average for the wood around and between the pins.

A further most important property of timber is that the cell walls become swollen on wetting, so that wet timber invariably occupies more space than dry timber. At moisture contents below fibre saturation value (approximately 28 per cent) the volume of timber varies approximately linearly with its moisture content. At moisture contents above saturation value, there is no variation since additional water occupies cell cavities where there is no swelling effect. The amount of moisture movement depends on the direction considered, typical average shrinkages for softwood between the saturated state and 12 per cent moisture content being as follows:

Tangential:	4–6 per cent
Radial:	2/3 of tangential
Longitudinal:	less than 0.1 per cent

These values reflect the fact that wood is most stable in the longitudinal direction, since most cells are aligned in this direction, while some stability in the radial direction is afforded by the rays. As would be expected, the moisture movement of timber varies considerably according to type. Perhaps surprisingly, within a given species, denser samples tend to shrink more than lighter samples and hardwoods shrink more than softwoods. An important consequence of the variation of movement of timber with direction is that distortion occurs when green timber is dried. Since tangential shrinkage is greater, growth rings tend to be thrown into tension and therefore become straighter on drying. Figure 8.6

shows exaggerated sketches of the effect of drying a number of sections cut from a log. It is noticeable that flat sawn boards tend to distort more than quarter sawn boards because their growth rings are longer. In the case of floors, flat sawn timber may also wear faster than quarter sawn equivalents because broader bands of softer springwood are exposed.

Seasoning

Seasoning is the controlled reduction of the moisture content of timber to a level appropriate to its end use.

Correctly seasoned timber should not be subject to further significant movement once in service, unless, of course, leaks occur. Seasoned timber is also immune to fungal attack, is stronger, has lower density and is therefore easier to handle or transport; it is also easier to work, glue, paint or preserve than wet timber. The importance of controlled reduction of moisture will be clear when it is appreciated that moisture loss always takes place at or near the surface of the timber, hence shrinkage occurs preferentially at these positions. Surface layers are therefore thrown into tension as they contract, resulting in 'fissures' in the form of checking (longitudinal surface rupture) or splitting (longitudinal rupture passing through the piece) if drying is carried out too quickly. Serious splits in large timbers are sometimes called 'shakes'. Evaporation tends to occur more quickly from the ends of timber during seasoning, so that checking and splitting are often more pronounced at ends.

Traditionally a large amount of timber was 'air seasoned' by stacking in open formations in a roofed enclosure. Drying takes place due to the effects of wind and sun, in a time which depends on the type and section size of timbers, the size of air gaps between them and the climatic conditions. The process minimises damage but is slow, often taking several years to complete, and will not reduce the moisture content below about 16 per cent, which is too high for most internal purposes. Its use in the UK has, therefore, largely been replaced by kiln seasoning in the case of softwoods, although it is still used for initial drying of hardwoods such as oak, which require a low rate of drying if damage is to be avoided.

Kiln seasoning involves heating the timber in sealed chambers, initially using steam to maintain saturation, the humidity then being progressively reduced to produce drying. Water has a lower surface tension and viscosity at higher temperatures and therefore flows more freely through the timber so that drying can take place more rapidly without damage. Drying schedules depend on the type and size of timber sections and their initial and target moisture content. Kiln seasoning takes about one week per 25 mm thickness for softwoods and about two weeks per 25 mm thickness for hardwoods. The process has the advantage that it sterilises the wood and can reduce moisture contents to lower levels than air seasoning – values of about 12 per cent can be reached quite easily.

8.4 Conversion of timber

This is the process by which the size of the log is reduced to a size appropriate for use and at which seasoning can be more easily carried out. A series of 'planks' is produced

Fig. 8.7 *Methods of conversion of timber: (a) 'through and through cutting'; (b) 'quarter sawn'.*

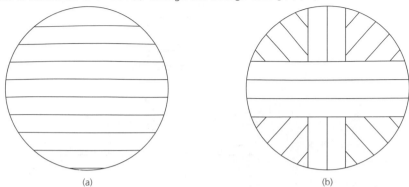

(a)   (b)

from each log, usually in the thickness range 25–75 mm. The cheapest and quickest way of doing this is by 'through and through cutting', as shown in Fig. 8.7(a). However, this results in some cut parallel to the grain ('flat sawn'), some cut at right angles to the grain ('quarter sawn') and some of sloping grain. All the non-quarter sawn timber will be subject to distortion on drying, particularly if relatively wide, thin sheets are required, though it may be argued that flat sawn timber, having a wider grain spacing, looks more attractive. An alternative is to cut the timber as in Fig. 8.7(b), though this is much more expensive, being slower and producing more waste than through and through cutting. Note from the figure that the flat-sawing process cuts the log in 10 stages and produces pieces of reasonable size, whereas the quarter-sawing process takes about 26 cuts in the saw mill and produces pieces of smaller average size, many of irregular shape.

8.5 Factors affecting the distortion of timber prior to use

These include the following (see Fig. 8.8 for an explanation of terms):

- **Species of timber**
  Some timbers that have similar moisture movements in the radial and tangential directions – for example, Douglas fir and utile – are inherently less prone to distortion than timbers that have high relative movements – for example, European redwood and beech.
- **Method of conversion**
  Quarter sawn timber distorts less on moisture change than flat sawn timber, though the latter has a more interesting grain pattern. Pieces of timber which contain the pith are also more prone to distortion, especially springing. Where the piece is not cut parallel to the longitudinal axis, bowing, springing and twisting are likely.
- **Slope of grain**
  Twisting is often caused when the grain is not straight or when the density of the wood varies – for example, due to unequal growth on different sides of the tree. Timber having uniform and straight grain is less likely to undergo distortion.

Fig. 8.8 *Types of distortion in timber and their measurement.*

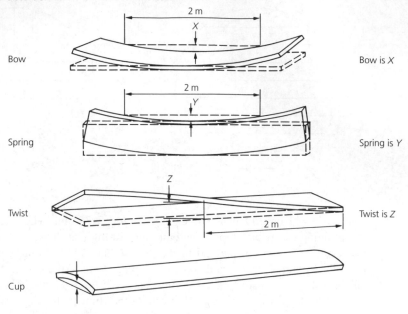

- **Stacking procedure**
  Each piece of timber in a stack must be adequately supported by 'stickers' prior to and during seasoning in order to avoid distortion due to self-weight. Twisting of timber can be largely avoided if it is held firmly in stacks during seasoning.
- **Moisture changes prior to installation**
  Rapid moisture changes after seasoning – for example, wetting or severe drying on occupation of new properties – can lead to serious distortion. Such changes should be avoided.

8.6 Durability of timber

Timber has remarkably good resistance to deterioration by atmospheric exposure – it is virtually unaffected by agencies such as rain, frost and acids, which are responsible for damage to other types of materials. The action of sunlight is only slight, prolonged exposure degrading the lignin adhesive so that the timber surface is bleached, becoming slightly fibrous. The main mechanisms of attack involve its being used as food – by plants in the form of fungi and by insects. In each case the destruction is brought about by enzymes which 'digest' the cellulose fibres and/or the lignin adhesive, converting them into food.

Fungal attack

Growth of all fungi in timber requires a moisture content of at least 20 per cent.

The enzyme action necessary for fungal growth cannot take place in completely dry timber. There are numerous varieties, some attacking the growing tree or the unseasoned green timber, while others are able to attack the wood in service. The former are mainly the concern of the timber producer, hence attention here is given only to fungi which are active in seasoned timber.

Where the **cellulose** only is destroyed, the wood breaks into small cubes and the rot is known as **brown rot**. Where both **cellulose and lignin** are destroyed, the wood becomes soft and fibrous, the rot being known as **white rot**.

Rots can also be divided into dry rot and wet rot. There are many species falling into the latter category but they are separated from dry rot because the remedial treatment of dry rot must be much more carefully carried out.

Dry rot

Dry rot (*Serpula lacrymans*, formerly called *Merulius lacrymans*) is a brown rot, so named because it finally leaves the wood in a dry, friable condition. The greatest risk is in areas without satisfactory ventilation such as cavities in floors or walls, especially where there are leaks or rising damp. Growth begins when rust-red spores come into contact with damp timber, the fungus developing as branching white strands (hyphae), which form cotton-wool like patches (mycelium) and finally soft, fleshy, spore-producing, fruiting bodies (sporophore). Once established, the hyphae can grow in adjacent timber, brickwork and plaster in search of moisture or food. They do not attack saturated wood but can transport excess moisture to drier areas of timber to continue the process. They can grow quite rapidly once established and may become modified into vein-like structures (rhizomorphs) 2 or 3 mm in diameter. Dry rot thrives at normal indoor temperatures but is killed at temperatures of over 40 °C – hence, it does not grow in positions subject to solar radiation. It becomes dormant at temperatures approaching freezing. Drying out of timber also renders the fungus dormant and it may die after one year in this state, though the process may take longer if the ambient temperature is reduced.

The fungus is more common in wetter parts of Great Britain, such as the north and west, though in any region it is most likely to occur in old derelict or damaged buildings. These are more often prone to dampness problems since older buildings are less likely to have the standard detailing found in modern properties – for example, cavities might be non-existent or may have become defective by bridging. Older properties in London have a high incidence of dry rot outbreaks, probably for this reason.

Dry rot can be a serious problem once established, since it is often very difficult to eradicate the hyphae completely, especially in brickwork or plaster or where access is limited. Normal remedial measures are:

- Cut out and burn all visibly affected timber, together with an extra 600 mm of apparently sound neighbouring timber which may contain hyphae.
- Remove any affected plaster, again working beyond the visibly affected area.
- Preserve sound exposed timber and treat surrounding brick or concrete with a fungicide.
- Replace affected timber with preserved, seasoned timber.
- Use zinc oxychloride in paint or plaster coatings applied to affected wall surfaces. This has a fungicidal action.
- Rectify the cause of the dampness.

Wet rot

As the name suggests, wet rots require higher moisture contents in order to thrive, optimum values being in the region of 50 per cent. There are many types of wet rot, most common in the UK being *Coniophora puteana* (formerly *Coniophora cerebella*) and *Phellinus contiguus* (alternative name *Porio contigua*).

Coniophora puteana ('cellar fungus') is commonly found in very damp situations in buildings – especially basements or cellars. The fungus is of the brown variety and leads to cube formation, especially in large timber sections, though a skin of sound timber may be present. Surface undulations in timber may be the first signs of damage. Affected wood turns dark brown or black but fruiting bodies are rare.

Phellinus contiguus is responsible for a considerable increase in window joinery decay which has occurred over the last 20 years or so. It is a white rot, hence wood breaks into soft strands. The cause of the increase in this type of rot is probably the use of sapwood in joinery, which, together with poor design and inadequate glues, leads to water penetration at joints. Since most joinery of this type is painted, the first signs of attack are often undulations in or splitting of the paint film.

Damage from these fungi is more serious in the southern half of Great Britain, due to the added effect of certain wood-boring weevils which are found in association with wet rots in these regions, increasing the rate of deterioration.

The essential difference in the remedial treatment of dry and wet rots is that the latter do not normally infect neighbouring dry timber, brickwork or plaster; hence it is necessary only to cut away and replace affected timber. It is nevertheless still important to remove the cause of dampness and preservation of new timber is recommended in case any temporary dampness should recur.

Insect attack

Insects, such as beetles that commonly infest timber in construction, have a characteristic life cycle: egg, caterpillar (larva), chrysalis and finally adult beetle. Beetles then emerge from the timber and fly to new timber to lay eggs and repeat the cycle, which may take several years. Most damage is done by larvae, which burrow through the wood leaving fine powder ('frass') behind. Flight holes are the normal visual means of locating infestations but quite advanced deterioration accompanied by serious strength loss may occur with only a small number of holes. Dust is often ejected from these holes and an indication of the degree of current activity can be judged from the colour of exit holes and dust produced – a lighter colour indicates a recent attack.

Insect attack does not require wet timber, though higher moisture contents are preferred and timber under the influence of central heating is unlikely to be attacked (though this does not remove the risk in unheated parts of buildings, such as loft spaces). Note that sapwood is much more susceptible to insect attack than heartwood. Attack is also less likely in wood products, such as chipboard, containing synthetic adhesives.

Table 8.2 shows some common types of insect. The **powder post beetle** is mainly a problem in timber yards, while the others can thrive in interior, seasoned timber. The seriousness of attack also varies according to species; for example, the **common furniture**

Table 8.2 _Common insect attacks on seasoned timber_

Beetle	Life cycle	Visual signs	Timber attacked
Common furniture (_Anobium punctatum_)	3–5 years; emergence May–August	Beetles 5 mm long, granular dust; 0.5 mm diam. flight holes	Sapwood of untreated hardwoods or softwoods
House longhorn (_Hylotrupes bajulus_)	3–6 years	A few large oval holes up to 10 mm diam.; surface swelling, beetles 10–20 mm in length	Sapwood of softwoods (Surrey)
Death watch (_Xestobium rufovillosum_)	4–10 years; emergence March–June	Large bun-shaped pellets; ticking during mating period, May–June	Mainly hardwoods, especially if old or decayed
Powder post (_Lyctus_ species)	1–2 years; emergence June–August	10 mm exit holes; fine dust	Sapwood of wide-pored hardwoods partly or recently seasoned

beetle is extremely widespread but normally takes many years to produce a noticeable number of flight holes or advanced deterioration.

The **house longhorn beetle** is relatively rare since it is quite sensitive to climatic conditions, but in parts of southern England, and particularly in West Surrey, attacks have been severe. Additionally, since the life cycle is up to 6 years and beetles leave a sound surface skin, serious damage can occur before many flight holes become visible. Roof timbers are mainly affected and Building Regulations require effective protection of roofing timbers to be used in certain local authority areas in these regions.

The **death watch beetle** is chiefly a problem in partially decayed oak timbers of older churches and other historic buildings and is often associated with dampness. Treatment in such cases can be difficult, due to access problems and the very large sections often involved.

8.7 Preservative treatments for timber (BS 1282)

Preservatives should be toxic to fungi and/or insects as required, be of sufficient chemically stability for their operational environment, be able to penetrate timber and be non-aggressive to surrounding materials.

For any given situation, the decision as to what (if any) preservative to use should be based on:

- The required lifetime of the component.
- The natural durability of the wood species being employed.

- The risk, in that situation, of damp or insect attack.
- The ease of inspection of the component.
- The consequences of attack; for example, whether failure of a component could lead to a safety hazard such as structural failure.
- The ease with which maintenance and repair work can be carried out.

Preservatives

These can be classified according to type:

Tar oil preservatives (BS 144)

Creosotes, the best-known examples, are derivatives of wood or coal, though coal distillates are superior. They are only suitable for external use, since they give rise to a noticeable colour and smell and render timber unsuitable for painting. They are, however, cheap and can be applied to timber with a fairly high moisture content. They tend to creep into porous adjacent materials such as cement renderings and plaster.

Water-borne preservatives (BS 4072)

These consist of salts based on metals such as sodium, magnesium, zinc, copper, arsenic and boron. Copper–chrome–arsenate (CCA) is a common example. They are tolerant of some moisture in the timber, are colourless, non-flammable, non-creeping and do not stain timber, which may be painted on drying out. In some varieties, such as CCA, insoluble products are deposited in the wood, resulting in excellent retention properties. Copper salts such as CCA colour the wood green and this is one way of confirming that the wood has been treated.

Although water-based preservatives are fairly cheap, they have the disadvantage that they swell the timber and will cause corrosion of metals in contact until drying out is complete – this taking about 7 days, depending on timber type and section thickness. Hence, they are best used as a preliminary protection followed by kiln drying, rather than as an in-situ treatment. Timber should be below fibre saturation point (28 per cent moisture content) before the application of water-based preservatives. If there is a risk of timber becoming damp in use, any embedded metals should be corrosion resistant; for example, galvanised (occasional dampness only) or stainless steel.

Organic solvent preservatives (BS 5707)

Examples are chloronaphthalenes, metallic naphthanates and penta chlorophenol. Penetration is excellent provided the timber is quite dry, hence they are often preferred where only brush or spray applications are feasible. They are non-creeping, non-staining, non-corrosive to metals, non-swelling and quick-drying. Painting is usually possible once drying is complete, which usually takes a few days. They are widely used for remedial treatment, though many organic solvents are highly flammable and present a fire risk until dried, especially when used in confined areas such as roof spaces. When used for new work the moisture content of the timber at the time of treatment should be that required for end use, since the preservative does not increase the moisture content. They are often used to preserve new joinery, though it is important that all machining/drilling, etc., be carried out prior to application if maximum protection is to be achieved. They are more expensive than the other types and release solvents into the atmosphere – a process which is increasingly being regarded as environmentally unacceptable.

Boron diffusion

In this process freshly felled timber is dipped into a concentrated borate solution which diffuses throughout the wood by means of its own moisture, hence the need to treat the wood in a 'green' state. The process can be used to advantage on 'difficult' timbers – those which do not readily admit preservatives.

Application methods

The technique of application is just as important as the choice of a suitable preservative, since, if it is poorly applied, it will fail to penetrate timber sections sufficiently. The difficulty increases with the size of timber sections, and some timbers such as Douglas fir have naturally low permeability to preservatives. Where machining or drilling operations are to be carried out, they should be completed prior to application of preservatives in order that the maximum penetration is obtained at all exposed surfaces. The following techniques can be adopted:

Brushing or spraying

These give only limited penetration of timber, although they may be the only feasible methods for in-situ remedial work. However, if several coats are applied and the treatment is repeated periodically, a useful degree of protection is achieved.

Dipping

This involves immersing the timber in the preservative for a period, which, if long enough, can give considerable penetration of preservative. In the 'open tank' method of dipping, the preservative is heated, then cooled, while the timber is submerged; this gives even better results. It is nevertheless not suited to hardwoods or low-permeability softwoods.

Double-vacuum process

A vacuum is applied to timber in a sealed chamber and the preservative then introduced. The vacuum is released, forcing the preservative into the wood. After a soaking period, the vacuum is reapplied to eject excessive preservative. This process is now widely used in conjunction with organic preservatives for external joinery such as window frames.

Pressure impregnation

This is often used for injection of water-based and tar oil-based preservatives. In the empty cell process, the air surrounding the timber is compressed, the preservative introduced and then a still higher pressure applied, forcing preservative into the wood. On releasing the pressure, excess preservative is rejected. If the timber is not pressurised prior to injection of the preservative (in some cases, a vacuum is first applied), the retention of preservative is increased (full cell method). This method is more expensive in terms of the quantity used and, since saturated, will take longer to dry. Good protection is nevertheless achieved.

8.8 Current preservation requirements for new building work

The Building Regulations (1991) give general requirements concerning the use of timber in situations which might cause dampness, together with specific requirements for timber used in areas in which there is a risk of attack by the house longhorn beetle.

Much more detailed requirements for timber to be used in specific situations in buildings is given in The National House Building Council Standard, in BS 5268: Part 5 for structural timber applications and in BS 5589 for a wider range of applications. Fundamental to such guides are knowledge of both of the natural resistance of timber to decay as well as the ease with which specific timber types can be treated.

The durability of timber is given the following classifications:

VD	very durable
D	durable
MD	moderately durable
ND	non-durable
P	perishable

In terms of resistance to treatment with preservatives, timbers are classified as follows:

ER	extremely resistant
R	resistant
MR	moderately resistant
P	permeable

Examples of classifications of some common timbers used if the UK are given in Table 8.3. It will be noticed from this table that hardwoods are generally (though not always) more durable and more resistant to penetration of preservatives than softwoods. The table applies to heartwood only.

Table 8.3 *Durability and treatability classification of heartwoods*

Durability class	Treatability class	Species	Hardwood (H) Softwood (S)
VD	ER	Iroko	H
	ER	Teak	H
D	ER	European oak	H
	R	Western red cedar	S
MD	R	Douglas fir (UK)	S
ND	R	European whitewood	S
	R	Sitka spruce	S
	MR	European redwood	S
P	P	Beech	H

The non-durable (ND) group referred to in Table 8.3 are all commonly used for structural timber in the UK. In the NHBC standard, they would be required to be preserved in the following situations:

- Where there is a risk of attack by the house longhorn beetle.
- In tiling battens.
- In all flat roof timbers.
- In timber used for external walls.
- In timbers in contact with the ground or below the DPC.
- In external woodwork.

It will be noticed that timber for use in roofs and floors is not normally required to be preserved against fungal attack. The risk in these situations is small, provided good construction techniques are employed.

European standards for preservation

Preservation requirements are in the process of being harmonised throughout the EU and a summary is given here to illustrate the rather different approach that will be adopted. The British Standards referred to above (for example, BS 5589) take a **prescriptive** or **recipe** approach to preservation.

As an example, if European redwood is to be used for window frames its natural durability and ease of preservation are first identified (see Table 8.3). They are MD and MR respectively for European redwood. BS 5589 gives preservation details for external woodwork above the DPC, which would include window frames. For example, if a 60-year life is desired, a copper–chrome–arsenate solution with prescribed full-cell process and pressure period would be acceptable. Durability is **assumed** to be obtained on the basis of a **prescribed** process.

Eurostandards operate by **performance** criteria. If the need for a preservative is identified, a level of **penetration** is specified together with a **retention level** for the preservative according to risk.

The greater the penetration of the preservative, the further a beetle or fungus would have to travel to reach an unpreserved core of timber while the greater the retention level, the greater will be the chance of complete mortality to the organism in the treated material.

Figure 8.9 shows how the preservative selection process works.

1. The hazard class is first defined (see Table 8.4). Additionally, the risk of insect attack must be assessed according to local knowledge.
2. BS EN 350–2 gives the natural durability of common species as follows:
   - Resistance to fungal attack on a scale of 1 to 5 (1 is best). The scale corresponds to the durability ratings given in Table 8.3:

Fig. 8.9 *Flow chart for determining preservative treatment required in a given situation (Eurostandards).*

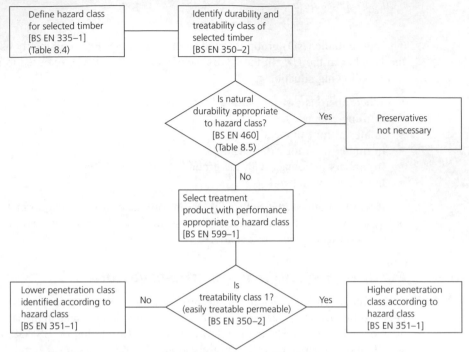

Table 8.4 *Hazard categories: definitions and examples*				
Data taken from BS EN 335–1				
Hazard category	Moisture content	Definition	Risk of fungal decay	Examples
1	< 20%	Under cover, fully protected from weather	Negligable	Internal joinery
2	Occasionally > 20%	As 1 but subject to high humidity (occasional wetting)	Low	Timber in pitched roofs; joists in ground floors
3	Frequently > 20%	Not covered; not in contact with the ground; continually exposed to weather	High risk	Cladding and external joinery
4	Permanently > 20%	In contact with ground-permanent weather exposure	Very hazardous; unacceptable risk	Fence posts, buried timber
5	Permanently > 20%	Permanently exposed to salt water	As 4 but marine environment	Marine piers, piling

        VD: 1
        D: 2
        MD: 3
        ND: 4
        P: 5

- Hardwoods are assumed to be resistant to the house longhorn beetle. Susceptibility of sapwood sections of softwoods to attack by the common furniture beetle or house longhorn beetle are indicated by:
    D   (durable)
    S   (susceptible)
    SH   indicates that heartwood is also susceptible.
- Treatability of timber must be assessed. A scale of 1 to 4 is given in BS EN 350–2 corresponding to the letter ratings of Table 8.3:
    P: 1
    MR: 2
    R: 3
    ER: 4

For example, European redwood is rated as follows (BS EN 350–2):
    Natural durability (fungi):   3–4
    Rating for all beetles: 'S'
    Treatability: 3–4   (heartwood)
                                   1   (sapwood)

3. The durability must be assessed against the hazard class identified (BS EN 460) in Table 8.5:

    o   indicates that natural durability against fungi is sufficient;
    x   indicates that preservatives are necessary;
    (o); (o)–(x); (x)   indicate intermediate cases where preservation may be necessary.

    One factor to be considered in intermediate cases may be the permeability of the timber – less permeable timbers probably absorbing less moisture on wetting and therefore being at lower risk of fungal attack. Sapwood of all species is regarded as durability class 5.

Table 8.5 *Assessment of need for preservation*

Depending on hazard class and natural durability

Hazard class	Durability class (fungi)				
	1	*2*	*3*	*4*	*5*
1	o	o	o	o	o
2	o	o	o	(o)	(o)
3	o	o	(o)	(o)–(x)	(o)–(x)
4	o	(o)	(x)	x	x
5	o	(x)	(x)	x	x

4. If a preservative is required the treatment product is selected according to hazard class and need for toxicity to beetles (BS EN 599–1). Products are subject to performance tests.

5. The penetration requirement of preservative depends on treatability of timber, penetration numbers P1 to P8 requiring progressively greater penetration according to hazard class (BS EN 351–1). For example, penetration level P3 requires 4 mm penetration laterally into sapwood (or into all wood if heartwood cannot be distinguished). For each hazard class an increasing level of retention of preservation is specified (R1 for hazard class 1, and so on up to R5).

Taking again the case of European redwood for window frames, Table 8.4 identifies window frames as hazard class 3. Heartwood has a natural durability rating of 3–4 (sapwood 5) (see page 237). Table 8.5 indicates ratings of (o) to (o)–(x) for this range of durabilities. On account of its low treatability (3–4) heartwood might be regarded as 'safe' to use without preservatives depending on the precise operating conditions and service life required. If sapwood is present, (o)–(x) rating, treatment would be recommended. (As indicated above, such timber would require a preservative if subject to NHBC regulations.) If preservation were to be used the penetration level P1 (low-permeability timber) and retention level R3 (hazard class 3) would be required.

8.9 Appropriate preservation levels

There are sound environmental reasons for keeping preservation requirements to a minimum in view of their fundamentally toxic nature and possible risks of pollution to plant, animal and human life at various stages. However there is also the need to protect property from timber deterioration. It is not surprising that the National House Building Council operate generally more stringent preservation requirements than Building Regulations since the former are primarily concerned about customer satisfaction and possible remedial costs to builders, whereas Building Regulations are principally concerned with health and safety. With the current widespread use of sapwood of softwoods it is probably appropriate that prescriptive preservation requirements are in widespread use, though in the longer term the environmental lobby may call for increased awareness of the risks of preservatives and the need to minimise risk of attack by:

- Appropriate design of buildings to minimise risks at source.
- Selection of naturally resistant timbers where possible.

There has been further concern over the use of sometimes drastic measures to combat fungal and insect attack in infected areas. The preservatives components of treated timber are of course non-renewable, since they become for practicable purposes a part of the wood. They cannot be extracted for reuse and materials such as creosote in particular may have implications for processing at the recycling stage if carried out.

Experiments have been carried out into control, particularly of fungal attack, by more environmentally friendly methods – for example, in the case of structural timbers in some older buildings as a part of conservation work. As indicated above, dry rot does not thrive in dry, well-ventilated environments so that it is possible, with careful monitoring, to destroy the fungus by controlling the environment. This also obviates the need to remove affected timber unless its strength is seriously affected. Perhaps the most important aspect of this type of protection is to carry out regular inspections; hence this technique is only feasible where timber is easily accessible for inspection or where the cost of a remote monitoring system is considered justifiable.

Rotting of window and door frames is one of the most common problems in building. The likelihood of this can be reduced by:

- Use of properly glued joints in joinery reduce the risk of cracking and water penetration at joints.
- Use of anti-capillarity groves (Fig. 8.10).
- Recessing frames from wall surface.
- Use of sealant around frame edges.
- Inward opening of external doors – these are then less vulnerable.
- Provision of weather bars to shed water clear of door bottoms.

Fig. 8.10 *Use of anti-capillarity grooves in window joinery.*

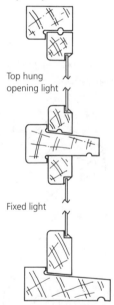

Top hung
opening light

Fixed light

8.10 Fire resistance of timber

'Burning' of wood occurs due to ignition of flammable gases, which are given off when the temperature of the material exceeds about 300 °C (pyrolysis).

There are two contrasting aspects of timber performance in fire:

1. The surface of wood soon heats up in a fire since the material has a low thermal conductivity. It therefore flames readily and in so doing
   - Spreads the fire by flame and heat emission.
   - Causes a hazard to occupants due to flame, heat and smoke emission.
   - May in some situations fail in a structural or other function.

   The speed at which this occurs depends on the section size and situation.

   Internal timber such as furniture is relatively dry and may burn quickly on account of quite small section sizes. In terms of the building itself, perhaps the most vulnerable components are tiling battens since these usually have very small sections. It is commonplace for tiled roofs to collapse for this reason in house fires.

2. In large structural sections the rate of burning is remarkably slow for the following reasons (Fig. 8.11):
   - Charcoal, like wood, is an excellent insulator (though charcoal itself burns at temperatures over 500 °C).
   - Gases produced by pyrolysis help to cool the charcoal as they diffuse to the surface before burning.
   - Moisture must be evaporated; some of this diffuses inwards, maintaining the moisture content at the core.

The charring rate of a flat timber surface is predictable and for fire resistance purposes is taken as:

Fig. 8.11 *Mechanism by which heat flow into timber exposed to fire is impeded.*

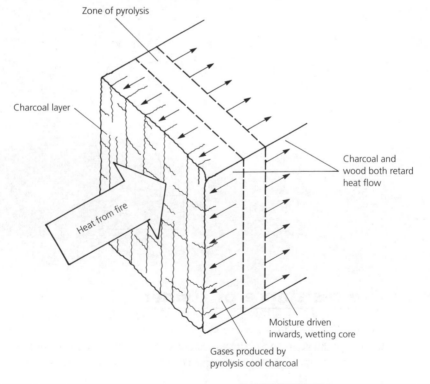

Zone of pyrolysis

Charcoal layer

Charcoal and wood both retard heat flow

Heat from fire

Moisture driven inwards, wetting core

Gases produced by pyrolysis cool charcoal

0.64 mm per minute (density less than 650 kg/m³)
0.50 mm per minute (density more than 650 kg/m³)

The lower rate applies to common hardwoods such as oak, utile and keruing. The higher rate applies to most softwoods, although a higher figure of 0.83 is indicated in BS 5268: Part 4 for Western red cedar. Rates may be increased when fire encroaches from more than one side.

It is therefore possible to design timber structures for fire resistance by provision of 'sacrificial' timber of thickness appropriate to the resistance required. In fact, since short-term strengths can be used for fire design purposes, it may, in any case, take up to half an hour in a serious fire before the short-term design stress is reached, larger sections being safer in this respect. Attention in structures will, however, need to be given to fixings – for instance, split ring connectors still transfer load if loosened by fire damage, whereas bolted joints may fail completely (see Fig. 8.12). In general, all metal fittings should be recessed beneath the char zone or protected from fire.

Fig. 8.12 *Structural glulam sections comprising curved (20 mm) laminates and straight (45 mm) laminates. The fixings are not fire protected – satisfactory for single-storey buildings only. All timber surfaces are coated with a clear fire retardant.*

It is found that laminated beams, which are usually bonded with phenol or resorcinol formaldehyde adhesives, have a fire resistance similar to that of solid timber of the same dimensions. Plywood box beams are less fire resistant than laminated or solid timber, since the thickness of wood is usually smaller.

Although timber floors might also be considered unsatisfactory, the normal first floor construction in houses has usually quite good fire resistance, largely due to its enclosed nature and the underlying plaster ceiling. In this respect, however, plywood is better than tongue-and-groove boards, which are better than flush boards, since a large number of joints or gaps increases ventilation.

In timber-framed buildings, satisfactory fire resistance can be achieved by the use of plasterboard on walls and ceilings, though firestops are also necessary in wall structures containing large vertical cavities.

Surface protection of timber from fire

A number of fire-retardant preparations are available. Some types impregnate the timber, increasing the proportion of charcoal and decreasing combustible gases in a fire. This treatment is carried out on seasoned timber cut to its final size. Typical chemicals include aqueous solutions of monammonium phosphate, borax and ammonium chloride, and preservatives are often included, though strength may be reduced by as much as 50 per cent due to cell wall damage if timber is not correctly dried afterwards. Alternatively, intumescent paints, varnishes or pastes may be used. These form protective films under the action of fire, preventing access of oxygen, and they do not adversely affect the strength of timber on treatment. Fire retardants are valuable in reducing surface spread of flame in cladding materials and partitions. Class 1 performance under BS 476: Part 7 can be obtained (class 3, untreated). Charring rates are, however, not greatly reduced and the best protection against structural failure is the provision of extra (sacrificial) timber to protect underlying material.

8.11 Timber for structural uses

The short-term tensile strength of perfect timber, taken in relation to its self weight, compares very favourably with that of steel, values of over 100 N/mm^2 being common in a material which has a density equal to about 7 per cent that of steel. The compressive strength of timber is much lower, due to the tendency of fibres to buckle on compression, but quite high bending strengths are nevertheless obtainable, timber being particularly useful for bending applications in situations where self-weight accounts for a substantial proportion of total weight – as in roofing and domestic flooring. The main problems in utilising timber efficiently for structural applications are as follows:

- Strengths, even of perfect ('clear') specimens, are very variable, coefficients of variation (equal to standard deviation divided by mean strength) of over 20 per cent often being obtained.
- *Long-term* strengths are much (typically 40 per cent) lower than *short-term* strengths, owing to the tendency of timber to creep.
- The strengths of individual pieces of bulk timber are greatly reduced by the presence of defects, such as knots and fissures, and by factors such as the rate of growth and slope of grain.

Considerable progress has been made in the last 15 years towards overcoming some of these difficulties by the introduction of techniques for grading each piece of timber according to its likely performance. Stress grading may be carried out visually or by machine.

Fig. 8.13 *Key terms relating to visual stress grading.*

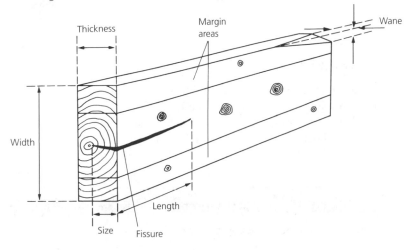

Visual stress grading (BS 4978)

Each piece of timber is examined visually for a number of defects which would limit strength, the timber hence being categorised into one of three grades: special structural (SS), general structural (GS) and reject. The SS grade is estimated to have a strength of between 50 and 60 per cent of that for clear (perfect) timber of the same species, and the (GS) grade approximately 30–50 per cent. Timber should be graded by a qualified grader and stamped with the grade and mark of a recognised authority. The Timber Research and Development Association (TRADA) is the main such body in the UK. (For experimental details, see Experiment 8.3 at the end of the chapter.)

The main factors to be considered are given below. (Key terms relating to visible stress grading are given in Fig. 8.13 and requirements for SS and GS grades are given in Table 8.6.)

Knots

The term 'knot area ratio' (KAR) is used to estimate the weakening effect of knots.

> *The knot area ratio is the proportion of any cross-sectional area which is occupied by knots.*

Since knots near to an edge have a more serious effect, a margin condition is defined as being present when more than *half* the top or bottom quarter of any section is occupied by knots (Fig. 8.14).

In BS 4978 a margin condition is described as

$$\text{MKAR} > \tfrac{1}{2}$$

The term 'TKAR' is then used to distinguish the overall KAR from the margin KAR. The key TKAR values are $\tfrac{1}{2}$, $\tfrac{1}{3}$ and $\tfrac{1}{5}$. Hence, the TKAR for the largest knot cluster for each piece must be placed within one of the following ranges:

$$TKAR < \tfrac{1}{5}$$

$$TKAR > \tfrac{1}{5}, < \tfrac{1}{3}$$

$$TKAR > \tfrac{1}{3}, < \tfrac{1}{2}$$

$$TKAR > \tfrac{1}{2}$$

KAR values in excess of one-third (margin condition) or one-half (no margin condition) lead to rejection of that piece. The KAR values for overlapping knots are added.

The process might seem time consuming but remember that the classification can usually be carried out quite quickly unless the KAR value is near to one of the above boundaries.

Table 8.6 *Visually stress-graded timber: properties and requirements*

Data taken from BS 4978

Property	Type	SS	GS
Knots	Margin condition (MKAR $> \tfrac{1}{2}$)	TKAR $\tfrac{1}{5}$ or less	KAR $\tfrac{1}{5} - \tfrac{1}{3}$
	No margin condition (MKAR $< \tfrac{1}{2}$)	TKAR $\tfrac{1}{3}$ or less	TKAR $\tfrac{1}{2} - \tfrac{1}{3}$
Fissures	Not through thickness	Not longer than half the length of the piece	Unlimited
	Through the piece	Not longer than twice the width of the piece	Not longer than 600 mm on any running metre
Slope of grain	—	Must not exceed 1 in 10	Must not exceed 1 in 6
Wane	—	Wane shall not reduce the full edge and face dimensions to less than two-thirds of the dimensions of the piece. Length of wane is unlimited	
Rate of growth	—	Average width of annual rings not greater than 6 mm	Average width of annual rings not greater than 10 mm
Distortion	Bow	Not greater than 10 mm over a length of 2m	Not greater than 20 mm over a length of 2 m
	Spring	Not greater than 8 mm over a length of 2 m	Not greater than 12 mm over a length of 2 m
	Twist	Not greater than 1 mm per 25 mm width over a length of 2 m	Not greater than 2 mm per 25 mm width over length of 2 m
	Cup	Unlimited	Unlimited

Fig. 8.14 *Effect of margin condition on TKAR requirements. (a) Since more than one-half of a margin area is occupied by knots, the knot area ratio must not exceed 1/3 for GS grade and 1/5 for SS grade. The TKAR for this section is about 1/2 – the piece is therefore 'reject'. (b) Since less than one-half of a margin area is occupied by knots, the knot area ratio must not exceed 1/2 for GS grade and 1/3 for SS grade. The TKAR for this section is between 1/3 and 1/2 – it corresponds to the GS grade.*

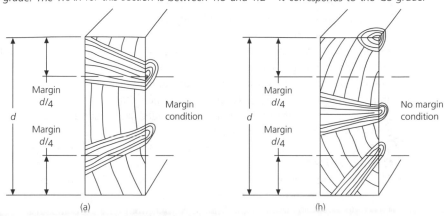

Fig. 8.15 *Planes of weakness resulting when log is not cut parallel to grain direction.*

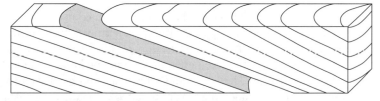

Fissures

These result from seasoning problems and are most common in larger sections, especially near the end of the piece. The size of a fissure is equal to the depth of cracks and the length of deeper fissures must be controlled in areas of higher bending moment, since they reduce shear resistance, and hence bending strength, parallel to the grain.

Slope of grain

This results where the tree does not grow straight or where the log is not cut parallel to the direction of growth. A plane of weakness is produced where the grain intersects the surface, leading to the possibility of tensile or shear failure (Fig. 8.15).

Wane

Wane occurs when a corner of the piece is cut too close to the outer circumference of the tree (Fig. 8.13). Its presence reduces bending strength in centre regions or bearing strength near the ends of a beam.

Rate of growth

It has been found that faster grown softwoods have lower strength and stiffness than slower grown examples of the same species. Growth rates depend on climatic and other conditions and often vary greatly during the life of the tree, but the rate of growth requirement guards against the use of excessively fast grown timber. (This rule does not apply to hardwoods.)

Distortion

The four types of distortion have already been described (Fig. 8.8), and several of these are often found in the same piece. However, some types are more probable and more serious than others for a given application; for example, cupping is found in floorboards and skirting boards, while bowing, springing and twisting often occur in structural timber. Bowing and twisting are a particular nuisance in wall plates, as are springing and twisting in hip rafters. It should be noted that, if timber is graded while wet and especially if recently removed from a banded pack, significant further distortion can occur afterwards during drying.

Other defects

These may include wormholes, resin pockets, sapstain and abnormal defects such as fungal decay or 'brittleheart'. See BS 4978 for guidance on such defects.

Machine stress grading (BS EN 519)

This is based on the correlation that has been found between bending strength (modulus of rupture) and modulus of elasticity, as determined by deflection tests on samples to bending (Fig. 8.16). A major attraction of the method is that it involves a simple mechanical test on each piece rather than relying on broad relationships between strength and measured defects. The 1 per cent strength value is obtained directly from the correlation of Fig. 8.16, the only additional factor required being the factor of safety. Figure 8.16 was

Fig. 8.16 *Correlation between elastic modulus and modulus of rupture for timber. (Crown copyright)*

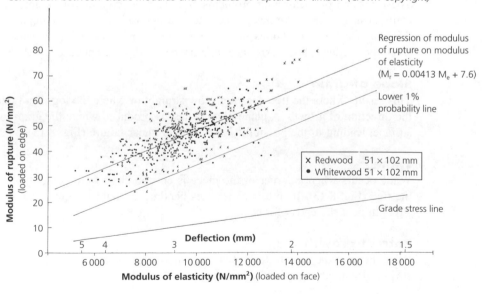

derived for European redwood (*Pinus sylvestris*) and European whitewood (*Picea abies*), which together form a very large proportion of the timber used in the UK at present.

Modern stress grading depends heavily on computer-operated machines which rapidly measure deflections under load of successive portions of each timber piece as it passes through. Timber is normally loaded on face rather than on edge for practical reasons and the machine automatically compensates for any out-of-straightness. Visual inspection is also carried out for defects such as fissures, wane, resin pockets, etc. Machine-graded timber is now classified directly into the appropriate BS 5268 strength class. An advantage of machine grading is that better grades are obtained than by visual grading of the same batch and there is less reject material.

Stress-grading machines must comply with an approving authority if grading in accordance with BS 4978, and machines are subject to unannounced periodic inspection in order to ensure that standards are maintained. Machine stress grading is now widely used for timber to be employed in structural elements, such as roof trusses and glued laminated timber.

Derivation of permissible stresses (BS 5268)

The safe working stresses for any particular piece of a given species have been derived by the following process:

1. BS 5268 stresses are based on strength tests carried out on grade SS dry timber of depth 200 mm. A special statistical distribution called the Weibull distribution was used to give failure strength levels in bending, with 5 per cent failures.
2. Grade stresses were obtained by multiplying the characteristic stress from (1) by factors allowing for duration of load, size and factor of safety. For bending, these factors are:

    (a)   Section depth factor: $0.849 = \left(\dfrac{200}{300}\right)^{0.4}$

    (Grade stresses are based on 300 mm sections, whereas experimental data is based on 200 mm depth sections.)

    (b)   duration of load factor: 0.563. This gives the long-term loading stress comparative to short-term test results in which creep is much reduced.

    (c)   General safety factor: 0.724.

    The factors (a), (b) and (c) combine to give an overall reduction of 0.346 (divide the characteristic stress by 2.89).
3. For GS timber, a relativity factor, obtained by experiment, must be applied to the SS value. For bending it is 0.71.
4. Other properties – for example, elastic moduli and compressive stresses – are obtained in the same way if experimental data is available on bulk samples. For some species, however, tests on 'clear' (perfect) samples, which used to form the basis of design, are still used in the absence of adequate data on bulk samples, conversion factors being based on known correlations between the behaviour of small clear sections and SS grade sections.

    To take an illustration for European redwood/whitewod, the SS grade characteristic bending stress, with 5 per cent failures, of a 200 × 100 mm section is 21.6 N/mm². Therefore the SS grade stress is:

$$\frac{21.6}{2.89} = 7.5 \text{ N/mm}^2$$

The GS grade stress will be:

$$7.5 \times 0.71 = 5.3 \text{ N/mm}^2$$

These stresses are seen to be well below the short-term bending strength of clear timber – typically 70 N/mm², the various causes of the large reductions having been outlined above.

Strength classes in BS 5268

The design procedure for structural timber can be quite complex, since the designer would need to be aware of the timber species available, together with stress grades for each species; it is not sufficient to refer to a stress grade such as 'SS' because different species have, in general, different stress capacities at a given grade. If, therefore, one species is no longer available, there may be some difficulty in finding a replacement with similar structural properties. The introduction of strength classes in the 1991 edition of BS 5268 helped to alleviate such problems by grouping different species/grades together, so that the designer need no longer specify an individual species (unless there are other reasons for such a requirement). Ten strength classes were defined, classes 1 to 5 generally being softwoods and classes 6 to 9 being denser hardwoods. Of these classes, SC3 and SC4 were common as they corresponded to European redwood and whitewood in the GS and SS grades respectively. The timber design tables in Building Regulations are based on strength classes SC3 and SC4 and form an extremely simple method of design for straightforward structures such as domestic floors and roofs.

In order to harmonise BS 5268 with the proposed EC5, the European code for the structural use of timber, new strength classes have been introduced with prefixes of C (or TR standing for 'trussed rafter') for softwoods and D for hardwoods. The numbers assigned correspond broadly to the characteristic bending strength of the timber in each class (about three times the permissible bending stress as explained above).

'Service classes' have also been introduced in anticipation of Eurocode 5. They are shown in Table 8.7 and are used in assessing allowable stresses for structural timber. They correspond broadly to:

- Internal heated (service class 1)
- Internal unheated (service class 2)
- External (service class 3)

The properties corresponding to the various strength classes are shown in Table 8.8. The figures given are valid for service classes 1 and 2. In the new system, European redwood and whitewood in the GS and SS grades would correspond to C16 and C24 respectively. Table 8.9 shows classification of some common species. The strength classification should be marked on timber as well as its stress grade, in order to simplify its selection for specific purposes. By grouping strengths into classes, some species will inevitably not be used at their full grade stress and the option remains to use selected species at their appropriate stress, though, for most purposes, the simplification provided by strength classes should be attractive to designers.

Table 8.7 *Structural timber: service classes and definitions*

Data taken from BS 5268

Service class	Definition
Class 1	Characterised by a moisture content in the materials corresponding to a temperature of 20 °C and relative humidity of the surrounding air only exceeding 65% for a few weeks per year. (Average moisture content not exceeding 12%)
Class 2	Characterised by a moisture content in the materials corresponding to a temperature of 20 °C and relative humidity of the surrounding air only exceeding 85% for a few weeks per year. (Average moisture content not exceeding 20%)
Class 3	Due to climatic conditions is characterised by higher moisture contents than service class 2

Table 8.8 *Structural timber: strength classes and properties*

Data taken from BS 5268 (all values N/mm^2)

BS5268 strength class	Bending parallel to grain	Tension parallel to grain	Compression parallel to grain	Compression perpendicular to grain (a)	(b)	Shear parallel to grain	Modulus of elasticity Mean	Min.
C14	4.1	2.5	5.2	2.1	1.6	0.6	6 800	4 600
C16	5.3	3.2	6.8	2.2	1.7	0.6	8 800	5 800
C18	5.8	3.5	7.1	2.2	1.7	0.6	9 100	6 000
C22	6.8	4.1	7.5	2.3	1.7	0.7	9 700	6 500
C24	7.5	4.5	7.9	2.4	1.9	0.7	10 800	7 200
TR26	10.0	6.0	8.2	2.5	2.0	1.1	11 000	7 400
C27	10.0	6.0	8.2	2.5	2.0	1.1	12 300	8 200
C30	11.0	6.6	8.6	2.7	2.2	1.2	12 300	8 200
C35	12.0	7.2	8.7	2.9	2.4	1.4	13 400	9 000
C40	13.0	7.8	8.7	3.0	2.6	1.4	14 500	10 000
D30	9.0	5.4	8.1	2.8	2.2	1.4	9 500	6 000
D35	11.0	6.6	8.6	3.4	2.6	1.7	10 000	6 500
D40	12.5	7.5	12.6	3.9	3.0	2.0	10 800	7 500
D50	16.0	9.6	15.2	4.5	3.5	2.2	15 000	12 600
D60	18.0	10.8	18.0	5.2	4.0	2.4	18 500	15 600
D70	23.0	13.8	23.0	6.0	4.6	2.6	21 000	18 000

Notes: 1. The higher compression perpendicular to grain figures (a) can only be used where specifications prohibit wane in bearing areas.
2. 'C' prefixes are for softwoods, 'TR' for trussed rafters and 'D' for hardwoods.
3. These figures apply to service classes 1 and 2.
4. Density requirements are also included in the standard.

Table 8.9 *Common timbers: strength classifications*

| Standard name | Strength class (Nearest 'C' equivalent) | | | |
	SC1 (–)	SC2 (C14)	SC3 (C16)	SC4 (C24)
Imported				
Parana pine			GS	SS
Redwood			GS	SS
whitewood				
Western red cedar	GS	SS		
Douglas fir-larch			GS	GS
(Canada/USA)				
Hem-fir (Canada)			GS	SS
British grown				
Douglas fir		GS	SS	
Scots pine			GS	SS
European spruce	GS	SS		
Sitka spruce	GS	SS		

Design of structural timber sections

This is may be covered by BS 5268 or Eurocode 5 which deal with a variety of structural components from solid timber as well as structures in glued laminated timber, plywood and hardboard. By way of illustration, the principles of design of solid timber joists and struts to BS 5268 will be considered.

Design of joists in simple bending

The species of timber to be used must first be identified, and the timber stress graded either visually or by machine. The most economical result will be obtained by using pieces of the same species and grade. Supposing, for example, that dry GS grade imported redwood is available. Table 8.9 shows that this timber is in strength class C16 and Table 8.8 shows that the timber has the following properties:

- Bending stress 5.3 N/mm^2
- Mean modulus of elasticity 8.8 kN/mm^2
- Minimum modulus of elasticity 5.8 kN/mm^2

Suppose that this timber is required to be used in a domestic floor of span 3 m in the form of joists at the normal spacing of 400 mm. It is necessary to estimate the load to be carried by the joists. The superimposed load would normally be the largest – in Building Regulations the value of 1.50 kN/m^2 applies in most cases. In addition, the self-weight of the floor must be calculated. For the purpose of calculation, the following might be assumed:

Size of joists: 50 × 200 mm

(The depth may not be correct but the error resulting will be small because the self-weight of joists is generally a small proportion of total weight.)

Thickness of floor boards: 20 mm

Density of timber: 600 kg/m^3

Weight of joists:

$$0.05 \times 0.20 \times 3 \times 600 \times 9.81 = 177\,\text{N}$$
(thickness)   (width)   (span)   (density)   (grav. accel.)

Weight of floorboards per joist

$$0.02 \times 0.4 \times 3 \times 600 \times 9.81 = 141\,\text{N}$$
(thickness)   (depth)   (span)   (density)   (grav. accel.)

(The mass of a ceiling, if used, would need to be included.)

Total self-weight = 177 + 141 = 318 N   or   0.318 kN

Total superimposed weight per joist = 1.50 × 0.4 × 3
                                               (loading)   (spacing)   (span)

$$= 1.800\,\text{kN}$$

Hence total load supported = 1.800 + 0.318 = 2.118 kN

For normal spans the bending stress limits the load which can be carried by joists. The bending moment (M) is given by

$$M = \frac{WL}{8} \qquad \text{where} \quad W = \text{load (UD)}$$
$$L = \text{span}$$

Working in newtons and millimetres

$$M = \frac{2\,118 \times 3\,000}{8}$$

$$= 7.94 \times 10^5\,\text{N mm}$$

Also,

$$M = fZ \qquad \text{where} \quad Z = \text{section modulus} \ (= bd^2/6)$$
$$b = \text{breadth}$$
$$d = \text{depth}$$

f, the bending stress, is 5.3 N/mm^2, though BS 5268 indicates that when four or more members act together such that the load is shared – as in boarded floor, the bending stress can be increased by 10 per cent (factor K_8). Hence, in this case, the permissible bending stress is 5.3 × 1.1 = 5.83 N/mm^2. Therefore,

$$Z = \frac{M}{f} = \frac{7.94 \times 10^5}{5.83}$$

$$= 1.36 \times 10^5\,\text{mm}^3$$

For $b = 50$ mm

$$d = \sqrt{\frac{6Z}{b}} = \sqrt{\frac{6 \times 1.36 \times 10^5}{50}}$$

$$= \sqrt{16\,300}$$

$$= 128 \text{ mm}$$

The standard size to cover this requirement would be 150 mm.

For larger spans, the deflection may be the limiting criterion and so this should be checked also:

$$\text{deflection (UD load)} = \frac{5WL^3}{384\,EI}$$

$$= \frac{5 \times 12WL^3}{384\,Ebd^3}$$

where E = modulus of elasticity

I = second moment of section = $\dfrac{bd^3}{12}$ (rectangular section)

Hence

$$\text{deflection} = \frac{60 \times 2\,118 \times 3\,000^3}{384 \times 8\,800 \times 50 \times 150^3}$$

$$= 6.02 \text{ mm}$$

(Note that E_{mean} is used here because load sharing is assumed. In other cases, or where the dead load forms a substantial proportion of the total load, E_{min} should be used.)

BS 5268 limits deflection of joints to 0.003 × span; in this case 0.003 × 3000 = 9 mm that the joint is satisfactory.

Note:
1. Exceptionally, where very short spans are used under high loads, the complementary shear stress may need to be calculated and compared with the permissible value given in BS 5268.
2. The bearing area at the ends of joists should be sufficient to ensure that the bearing stress in the timber (compression perpendicular to grain) does not exceed the value allowed. See Table 8.8.

Design of struts

This is somewhat more complex than the design of joists because the most likely form of failure of a strut is by buckling, a phenomenon which depends very much on the context of use. In particular, the size of strut needed will depend on the following factors as well as on the axial load applied:

- The length of the strut since longer struts are more likely to buckle.
- The degree of restraint at the ends. The following situations are possible:

- *No restraint* – for example, if the strut rests on a flat surface.
- *Positional restraint* – for example, in a small socket.
- *Directional restraint* – for example, built into brickwork with substantial embedment of 300 mm or more.
- The application of eccentric loads or bending moments to the strut; for example, if a floor joist is supported on one side of the strut.

In order to illustrate the principle of design, an example will be taken of a strut in the form of a stud partition, Fig. 8.17.

Fig. 8.17 *Design of struts in a stud partition.*

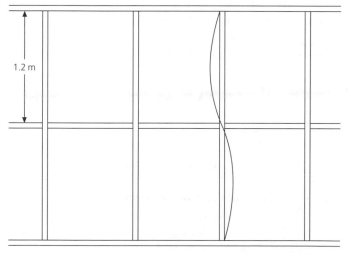

It is estimated that an axial load of 10 kN is likely to act on each strut in this partition and the common section size of 100 × 50 mm will be tried using grade C24 timber. The studs should be considered in two respects:

(a) As struts of length 1.2 m and thickness 50 mm, restrained in position (not direction) at each end in the horizontal direction in the plane of the paper. In case it is considered that the centres of the struts are restrained in direction. A possible mode of failure is illustrated in Fig. 8.17.

(b) As struts of length 2.4 m and thickness 100 mm, restrained in position (not direction) at each end in the horizontal direction at right angles to the plane of the paper.

First the effective lengths of the strut can be found from the values given in Table 8.10 (BS 5268).

The effective lengths in the case of these struts are the same as the actual length: 1.2 m and 2.4 m in the two directions respectively. The tendency to buckling of a strut depends on its 'slenderness ratio', which is the ratio of the effective length as defined above to the radius of gyration about the axis in question. For a rectangular section there is a simple relationship between radius of gyration and thickness, so that the ratio

Table 8.10 *Effective lengths of timber compression members*

Data taken from BS 5268

End conditions	Effective length/Active length
Restrained at both ends in position and direction	0.7
Restrained at both ends in position and one end in direction	0.85
Restrained at both ends in position but not direction	1.0
Restrained at one end in position and direction and at the other end in direction but not position	1.5
Restrained at one end in position and in direction and free at the other end	2.0

$$\frac{\text{effective length}}{\text{thickness}}$$

can be used as an alternative. In this case the effective lengths are

$$\frac{1\,200}{50} = 24 \text{ in the plane of the paper}$$

and

$$\frac{2\,400}{100} = 24 \text{ at right angles to the plane of the paper}$$

Hence, in this case there is equal stability in each direction. Also affecting performance is the strain at which the strut is being expected to work – higher strains lead to an increased risk of instability. BS 5268 uses the inverse of this ratio which gives numbers that are easier to work with. Table 8.11 gives modification factors (K_{12}) by which the grade stresses given in Table 8.8 for compression parallel to the grain must be multiplied when designing struts. In this example, for C24 timber, the grade stress is 7.9 N/mm^2. The elastic modulus for this strength class is 7.2 kN/mm^2 (the *minimum* value must be used unless there is load sharing between the struts). The ratio of elastic modulus to grade stress for this material is therefore

$$\frac{7\,200}{7.9} = 911 \quad \text{(no units)}$$

The two values 24 and 911 can now be referred to the X and Y axes, respectively, of Table 8.11. To the nearest 0.01 the table gives 0.50; hence the applied stress must not be greater than

$$0.50 \times 7.9 \text{ N/mm}^2 = 3.95 \text{ N/mm}^2$$

Table 8.11 Modification factor K12 for timber compression members

Not for design purposes (data based on BS 5268)

E	Ratio of effective length to breadth (rectangular sections)										
Grade stress	1.4	3.0	6.0	12.0	18.0	24.0	30.0	36.0	42.0	48.0	54.0
600	0.97	0.95	0.90	0.77	0.58	0.41	0.29	0.21	0.16	0.13	0.10
800	0.97	0.95	0.90	0.78	0.63	0.48	0.36	0.26	0.21	0.16	0.13
1 000	0.98	0.95	0.90	0.79	0.66	0.52	0.41	0.30	0.24	0.19	0.16
1 200	0.98	0.95	0.90	0.80	0.68	0.56	0.44	0.34	0.27	0.22	0.18
1 400	0.98	0.95	0.90	0.80	0.69	0.58	0.47	0.37	0.30	0.24	0.20
1 600	0.98	0.95	0.90	0.81	0.70	0.60	0.49	0.40	0.32	0.27	0.22
1 800	0.98	0.95	0.90	0.81	0.71	0.61	0.51	0.42	0.34	0.29	0.24
2 000	0.98	0.95	0.90	0.81	0.71	0.62	0.52	0.44	0.36	0.31	0.26

The applied stress is

$$\frac{10\,000}{100 \times 50} = 2.0\ \text{N/mm}^2$$

This is well within the allowed stress so that the struts in the partition should be adequate.

Eurocode 5

In terms of material, Eurocode 5 contains an important additional consideration for design. Whereas in BS 5268, creep in timber is allowed for by limiting long-term stress, Eurocode 5 gives additional requirements for long-term deflection levels. Creep deflection (k_{def}) factors must be applied to calculated deflections to give long-term deflections which must then not exceed stated limits. The k_{def} factors are stated as proportions of initial elastic deflections to be applied in addition and they depend on service class (Table 8.7). Timber to be used in damp situations must have a much bigger creep allowance. Where long span beams are used, creep deflection may be the critical factor in design.

8.12 Timber products

Most timber products are the direct consequence of the excellent bonding properties of the material. Smooth flat surfaces are easily obtained by planing and present a porous surface into which adhesives can penetrate. The adhesive therefore grips the wood, providing bond strengths that are often in excess of the cohesive strength of the material.

Timber products may offer the following advantages over solid timber:

- They are competitive in price on account of lower wastage rates. Smaller sections of the tree, together with defective pieces and off-cuts, which are not suitable for solid timber, are often perfectly acceptable for timber products.
- Products can be produced in much larger sizes than could be obtained in solid timber – for example, sheets or long lengths.
- Defects associated with solid timber can be removed or made less significant.
- Weaknesses and distortion associated with grain can be at least partly overcome.

Areas in which timber products could cause problems are:

- The product can only be as good as the adhesive used in its production. Some products deteriorate quickly in damp conditions – as might prevail on a building site during construction, or in areas such as kitchens or bathrooms.
- Some adhesives increase wear and tear on cutting tools used for processing products.
- Some timber products are not as aesthetically attractive as their solid timber counterparts.
- Bending strengths are not in general as good as those of solid timber parallel to the grain.
- Creep levels may be increased.

Sheet materials

Examples of common sheet products are shown in Figure 8.18 and a selection of properties is shown in Table 8.12. These are examples only of types obtainable and permit a broad comparison of the range of properties obtainable. The operating environments 'D', 'H' and 'E' broadly correspond to the hazard classes 1, 2 and 3 shown in Table 8.4, but manufacturers' recommendations should always be followed. The bending strengths give a guide to flexural loads that can be applied to sheets. The internal bond strengths relate to the cohesive strength of the materials. The elastic moduli give a guide as to what deflections can be expected.

Fig. 8.18 *Examples of common sheet materials. From the left: Douglas fir shutter ply; oriented strand board (OSB), tongue and grooved; particleboard (tongue and grooved); medium-density fibreboard (MDF) cornice moulding; medium board; hardboard.*

Table 8.12 Common sheet materials: properties and codes

The codes should be used to identify and check performance levels of each sheet

Material	BS EN number	Dry (D) Humid (H) External (E)	Code	Typical thickness (mm)	Bending strength (dry) (N/mm²)	Internal bond (dry) (N/mm²)	Elastic modulus (kN/mm²)
Hardboard							
(gen. purpose)	622–2	D	HB	4	32	0.6	–
(gen. purpose)		D/H	HB.H	4	32	0.6	–
(gen. purpose)		D/H/E	HB.E	4	35	0.6	–
(loadbearing)		D	HB.LA	4	32	0.6	2.5
Medium board							
(gen. purpose)	622–3	D	MBH (h.den. version)	10	15	0.1	–
(gen. purpose)		D/H	MBH.H	10	18	0.3	–
(gen. purpose)		D/H/E	MBH.E	10	21	0.3	2.4
(loadbearing)		D	MBH.LA1	10	18	0.1	1.8
Softboard							
(gen. purpose)	622–4	D	SB	15	0.8	–	–
(gen. purpose)		D/H	SB.E	15	1.0	–	–
(gen. purpose)		D/H/E	SB.E	15	1.1	–	–
(loadbearing)		D	SB.LS	15	1.1	–	0.13
MDF							
(gen. purpose)	622–5	D	MDF	10	15	0.6	2.5
(gen. purpose)		D/H	MDF.H	10	26	0.8	2.5
(loadbearing)		D	MDF.LA	10	27	0.65	2.8
Particle board							
(gen. purpose)	312–1	D	P2	18	11.5	0.24	–
(loadbearing)	312–4	D	P4	18	15	0.35	2.15
(loadbearing)	312–5	D/H	P5	18	16	0.45	2.4
(heavy duty)	312–6	D	P6	18	18	0.5	3.0
OSB							
(gen. purpose)	300	D	OSB/1	18	16/8	0.26	2.5/1.2
(loadbearing)		D	OSB/2	18	18/9	0.30	3.5/1.4
(loadbearing)		D/H	OSB/3	18	18/9	0.3	3.5/1.4
(heavy duty)		D/H	OSB/4	18	26/14	0.4	4.8/1.9

Wood fibre boards

These are produced by breaking wood down into its fibres and then pressing them together. The product obtained depends on the process used (wet or dry), the adhesive used (if any) and the pressure applied. They all have smooth surfaces reflecting their fibrous nature (though hardboards have a textured back surface produced by a wire mesh against which they are pressed to extract water during manufacture). Products are classified by BS EN 316.

Wet process
Hardboards (HB)   Density greater than 900 kg/m³
Medium boards: high density (MBH)   Density between 560 & 900 kg/m³
                low density (MBL)   Density between 400 & 560 kg/m³
Softboards (SB)   Density less than 400 kg/m³

Dry process
Medium density fibreboards (MDF)   Density greater than 600 kg/m³

Hardboards

These are characterised by high densities, bending strengths and bond strengths. These properties can be compared with those of other products in Table 8.12. Standard types must, however, be kept dry. They are widely used for internal linings for floors (under sheet coverings) and door skins. They are usually conditioned by moisture application to the rear surface to avoid slight rippling in use. Moisture-resistant types can be used for external claddings. Hardboards represent a low-cost, efficient use of timber in situations for which they are suitable. They have even be used, fixed to a timber frame, to carry loads over small internal openings ('stressed skin' construction).

Medium boards

These can be used where low-cost linings with better acoustic or thermal insulation are required. They can also be plastered.

Softboards

These are more suitable for ceiling applications where abrasion is not a problem. They have good sound absorption properties.

MDF boards

These boards incorporate a synthetic adhesive and are produced using heat and pressure. They are in general slightly weaker than hardboards and standard types must be kept dry. Smooth, high-quality surfaces are obtainable and for this reason they are often shaped and surfaced with a thermoplastics skin to produce substitutes for wood mouldings, with good dimensional stability at low cost. There are no grades recommended for external (hazard class 3) use.

Particle boards (chipboards)

These are formed by compressing small particles together under high pressure. The process results in sheets with a fine, smooth surface, but a coarser core. Hence they should not be planed as the surface quality would be lost. Grades rated as 'D' in Table 8.12 will undergo irreversible swelling if subjected to prolonged moisture so that they should not be used in building at construction stage until all risk of wetting is past. Loadbearing versions (P4) of these materials in tongue and grooved form are excellent for flooring, producing a flat smooth surface at a fraction of the cost of solid timber. For bathroom or kitchen areas moisture-resistant forms (P5) are required. Chipboards make excellent decking materials provided moisture-resistant forms are employed. Types with a bitumen precoating are available for flat roof applications. Creep levels in chipboard tend to be higher than those in solid wood, especially in humid atmospheres. There are no grades recommended for external (hazard class 3) use.

Oriented strand board

These are the latest type of board to be marketed. They comprise slivers of wood up to 75 mm long and about half a millimetre thick, bonded together under high pressure, typically with a waterproof adhesive. The slivers are aligned along the length of the board and the bending and stiffness figures in Table 8.12 show longitudinal and transverse performance. The table shows that in the longitudinal direction the boards perform better than similar chipboards, though they are more expensive. Tongue and grooved forms are available for flooring and the material can also be used for roofing and sheathing applications. There are no grades recommended for external (hazard class 3) use.

Plywoods

Plywoods are obtained by cutting rotary veneers from timber logs and bonding them together, alternate layers being crossed. In order to balance stresses due to movement, an odd number of plies is preferred. Plywoods therefore have good strength properties in both directions with better nailability near the edge than other sheet products. Very high quality veneers such as Finnish Birch, combined with high-performance adhesives, lead to excellent long-term stability in demanding situations such as eaves soffits. Coarser surfaces, which may include fine splits and repairs to knot defects, can be used for applications such as shutter ply. Such materials are much more expensive than sheet products such as particle board and OSB but they should be capable of more reuses.

In general, plywood is tending to be replaced by the cheaper alternatives above for all but very high-performance applications, due to the high cost of the quality timber required.

Glued laminated timber (glulam)

This comprises laminates of typically 45 mm thickness (straight types) or 20 mm thickness (curved types) bonded by synthetic adhesives to form members far larger than could be obtained from solid timber (Fig. 8.12). In fact the only limitations of size for glulam are transporting considerations. Members over 50 m in length can easily be produced. Finger jointing, combined with staggering of joints in individual laminates, produces full structural continuity throughout the material. An important feature of glulam is that because differences between individual pieces are less important than when used individually, greater stresses may be permitted. Poorer grades of timber can also be located at the core of the material where stresses are lower. Important aspects of good performance include:

- Good-quality smooth flat surfaces are essential for good bonding.
- Timber must be dry (less than 18 per cent moisture content) at the point of assembly.
- Waterproof adhesives are always recommended in case exposure to moisture occurs. Resorcinol or melamine formaldehyde adhesives are normally used. See BS EN 301 types I (external) or II (internal).
- It is good practice to keep exposed timber well clear of the ground (by supporting on concrete piers, for example) and to protect adequately from sun and rain.

Like solid wood, glulam is subject to creep (i.e. increased deflections caused by stress over a long period of time). The level of creep will depend upon the service conditions, and experience shows that if glulam gets wet at the point of construction creep levels may be much higher. Similarly, where the material is exposed to high relative humidities and, in particular, deep humidity cycles in use, creep levels will rise. Creep levels will be reduced if the timber is well protected from such humidity changes. Fire protection in the form of intumescent coatings will normally be required and this should also serve to reduce creep levels. In normal environments the long-term creep deflection of glulam beams is approximately equal to the initial elastic deflection.

The use of glulam has been traditionally limited to roofs of single-storey buildings where self-weight of the roof forms the main loading on members. There is, however, scope for use of this environmentally attractive material to be extended to multi-storey construction, of, for example, up to 5 or 6 storeys.

8.13 Environmental considerations in the use of timber

This chapter began with a description of the environmental advantages of timber, including its low energy input and the fact that a growing tree is a means of controlling or even reducing levels of carbon dioxide in the atmosphere. There have nevertheless been criticisms that the construction industry is at times indiscriminate in its use of timber, some hardwoods in particular being derived from mass de-forestation projects which may, in addition to carbon dioxide implications, lead to:

- Land instability
- Soil erosion
- Climatic changes on a local or wider scale
- Disturbance of wildlife balance

Many hardwood woodlands have long since disappeared from the UK and large parts of America and Africa. Current sources are often more distant and in less developed parts of the world, where economic and political influences are often of short-term rather than long-term focus. Clearly, the continued exploitation of natural resources on this scale may have serious long-term consequences and efforts are being made to ensure that timber supplies are obtained from **sustainable sources**. The term 'sustainable' was defined in Chapter 1 and in this context relates to adverse environmental consequences of extraction of timber. Since these can be very difficult to estimate at the time of extraction, the preferred term in current use at the moment is 'well-managed' sources.

Although there are no current regulations concerning the operation of timber importers there are now some international guidelines concerning identification of well-managed sources as well as independent accreditation organisations to ensure that suppliers claiming compliance with guidelines do, in fact, practise their declared procedures. However, world-wide uniformity has not yet been achieved and not all timber exporters operate to such schemes. Many importers are now taking steps to ensure that hardwood imports are from well-managed sources, though the point has already been reached where certain timbers are threatened with extinction. Examples of these are species within the groups:

- African walnut
- Afrormosia
- Rosewood
- Ebony
- Mahogany
- Ramin
- Utile

Such timbers should be regarded as particularly environmentally sensitive. As there are some botanical species within these groups that are not under threat, a total ban would not be appropriate. Nevertheless, to avoid risk of further depletion it would be wise to consider alternative, perhaps lesser known, species with similar properties. More flexible specifications permit the use of alternative species, often from well-managed forests, and this reduces the pressure on well-known species from unsustainable sources that are under threat.

Recycling timber

It will be recalled from Chapter 1 that the preferred order in relation to recycling of any product is:

Reduce – Reuse – Recycle – Recover

Reduce

There are opportunities, particularly in factory production, for waste to be reduced. For example, some joinery sections are now produced by finger jointing them into continuous lengths prior to cutting to size, thereby eliminating offcuts. This technique assumes, of course, that the finger joint is equivalent in strength and durability to solid wood, and this is a quality control issue. While the proportion of offcut waste in on-site work can be reduced by careful ordering, it is much more difficult to eliminate waste completely in this context.

Reuse

Demolition sites can be a useful source of quality timber. Most demolition work is in older buildings and it is found that timber that has already been in service for a century or more can be of very sound quality. In fact, such timber may well be heartwood of mature trees and perform better than modern softwood equivalents which often contain sapwood. Such timber should, of course, be checked for insect or fungal damage. Good candidates for reuse are floorboards and parquet flooring. These are often of high quality and can be used with minimal preparation if used upside down. The reuse opportunities for timber for structural purposes are limited unless it is subjected to stress grading, and this is in general not feasible. Reuse applications are currently therefore of limited scale and operated on a local largely unstructured basis.

Recycle

Bearing in mind the nature of some of the timber products referred to above, and particle board in particular, recycling would seem to be a very attractive option for timber construction waste. However, to maintain quality and a 'new' appearance, particle board manufacturers tend to have stringent specifications for wood chip waste. Recycled wood chip contents are currently limited to around 5 per cent, so low that there is considerable overcapacity in the market. Perhaps there is a need for a lower grade board for nominal applications which could accept larger quantities of waste material. It will be appreciated that some contaminants such as paint, adhesives and nails can be quite difficult to remove from wood.

A further possibility for low-grade material is to form it into insulants. Figure 8.19 shows an insulant, made from low-grade wood chips and hydrated lime, being used as an in-situ insulant in the walls and floors of a timber-framed dwelling.

Fig. 8.19 *Use of lime-bonded wood chips as an insulant in a suspended ground floor.*

Recovery

This constitutes the least preferred option in that the timber is used only as a fuel. Nevertheless, it must be considered preferable to use the timber to assist in, for example, district heating schemes from combustion of waste materials, rather than simply burning the material in the open. Large bonfires are a common site on demolition sites and must be regarded as lost environmental opportunities. There have been examples of processes in which timber is pelletised for burning in special wood-burning stoves, though such appliances are rare at present in the UK.

Experiment 8.1 Identification of common woods

Identification of a given piece of wood, especially if a softwood, can be extremely difficult since there are many hundreds of species obtainable and properties within each species often vary. The following may nevertheless assist as a guide to identifying some of the commonly available softwoods and hardwoods.

It is usually essential to have a smooth surface sample of the wood to be identified – this could be in the form of a small rectangular piece. One radial and one tangential face should be planed smooth and one transverse section cut smooth with a **very sharp** knife or chisel. With careful cutting a clean, dark surface, free of tearing, should be obtainable. Sanding will also produce a smooth surface but this takes longer and the dust tends to obscure finer detail such as resin ducts. Moistening the end of the timber may help to find these. Low-power (e.g. ×10) magnification will assist greatly in seeing and identifying detail in transverse sections. It may also be an advantage for some assessments such as knot distribution to have a larger piece of timber as well as the smaller sample.

A distinction should first be made between hardwoods and softwoods. On the transverse plane, softwoods can be recognised by the absence of pores – for example, Fig. 8.20 (Douglas fir). Hardwoods can be further subdivided according to the nature of pores, whether ring porous or diffuse porous (Figs 8.3 and 8.23). The density of

Fig. 8.20 *Transverse section of Douglas fir (Pseudotsuga taxifolia) – softwood. Note dense summerwood bands. (×8 magnification)*

Fig. 8.21 *Transverse section of whitewood (Picea abies) – softwood. Note the small, light-coloured resin ducts irregularly spaced in latewood. (×8 magnification)*

Fig. 8.22 *Transverse section of beech (Fagus sylvatica) – diffuse porous hardwood. Note prominent rays. (×8 magnification)*

Growth ring

dry specimens should also be measured by weighing and dividing by volume. Tables 8.13, 8.14 and 8.15 give characteristic descriptions of the key properties of common softwoods, ring-porous hardwoods and diffuse-porous hardwoods, respectively. Figures 8.20–8.23 show transverse sections of a number of common woods and these may assist in identifying the species concerned. Reference to scanning electron microscope photographs of Figs 8.3 and 8.4 may also be helpful.

Fig. 8.23 *Transverse section of African mahogany (Khaya ivorensis) – diffuse porous hardwood. Note the well-spaced pores and thin rays. The growth rings are indistinct in this sample. (×8 magnification)*

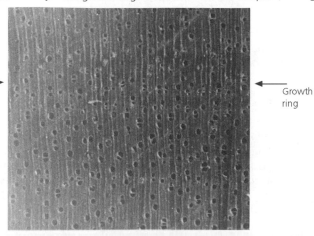

Growth ring

Table 8.13 *Common softwoods: properties and identification*

In approximate order of darkening colour

Type	Growth rings	Density (kg/m³)	Colour	Other features
European whitewood (European spruce) *Picea abies* (Fig. 8.21)	Well-defined; springwood merges gradually	510	Very light yellow/white	A few resin ducts irregularly spaced in late wood
Sitka spruce *Picea stichensis*	Springwood merges gradually into thin bands of dark red/brown summerwood	400	Very light brown	Irregular resin ducts; latewood; rapid growth; rings well spaced; low density
European redwood (Scots pine) *Pinus sylvestris* (Fig. 8.3)	Well defined; thin/ medium bands of dark red summerwood	520	Sapwood light brown; heartwood deeper red/ brown	Newly cut surface smells resinous; regular resin ducts; knots in groups separated by clear timber

Table continues

Type	Growth rings	Density (kg/m³)	Colour	Other features
Parana pine *Araucaria augustifolia*	Poorly defined; early wood merges gradually into latewood	530	Sapwood light; heartwood light to dark brown with red streaks	No resin ducts; dense, fine, straight grain
Western hemlock *Tsuga haterophylla*	Abrupt transition from springwood to pink/brown summerwood	500	Yellow/brown	Resin ducts absent
Douglas fir *Pseudotsuga taxifolia* (Fig. 8.20)	Well defined springwood; quite thick well-defined bands of dense hard red/brown summerwood difficult to cut; growth rings often wavy	545	Darker than most softwoods except cedar; sapwood yellow; heartwood red/brown	Resin pockets visible on transverse section under microscope; may be resinous smell
Western red cedar *Thuja Plicata*	Clear rings; narrow bands of latewood	370	Sapwood yellow; heartwood fairly dark brown	Distinctive aromatic smell; no resin ducts; note the low density

Table 8.14 *Common ring-porous hardwoods: identification*

Type	Growth rings/rays	Density (kg/m²)	Colour	Other features
Ash *Fraxinus excelsior*	Growth rings distinct; fairly course pores in springwood; hard, dense summerwood; rays invisible	690	Very light-whitish	Quite hard, dense wood
Oak *Quercus robus* (Fig. 8.4)	Rays very clear; large pores in springwood	690	Light yellow/brown	'Silver grain' on quarter-sawn surfaces (effect of rays); hard, dense
Sweet chestnut *Castanea sativa*	Well-defined thin rings of large pores; rays indistinct	600	Light/medium brown	Similar to oak if flat sawn; silver grain absent when quarter-sawn; less dense that oak

Table 8.15 *Common diffuse-porous hardwoods: identification*

Type	Growth rings/rays	Density	Colour	Other features
Beech *Fagus sylvatica* (Fig. 8.22)	Thin, well-defined rays on transverse section; grain fairly faint; pores not visible except under microscope	710	Light yellow/ pink	Rays produce fine, well-distributed lines on flat sawn section
Maple *Acer saccharum*	Summerwood distinct though only slightly darker than springwood; rays distinct; fine, well-scattered pores	690	Light biscuit brown	Fine smooth surface; attractive lustre; hard, dense
African mahogany *Khaya ivorensis* (Fig. 8.23)	Well-shaped medium-size pores; thin rays – clearly visible under microscope	660	Dark red/ brown	Pores produce long fine indentations on radial and tangential surfaces

Fig. 8.24 *Methods of cutting timber for moisture movement measurement.*

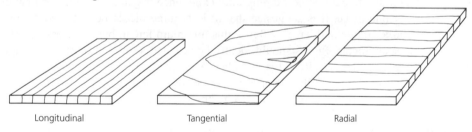

Longitudinal   Tangential   Radial

Experiment 8.2   Measurement of the moisture movement of timber

This can be quite easily carried out, provided suitably sized specimens are available. Ideally, three pieces of timber of each species should be available of size approximately $6 \times 75 \times 100$ mm, each cut with its long dimension parallel to one of the three respective principal directions in the wood (Fig. 8.24). Knot-free samples are preferable and the end grain of the flat-sawn piece should be as straight as possible if cupping is to be avoided. Pieces containing the pith should, in any case, be avoided.

Note that quarter-sawn specimens are fragile and must therefore be handled with care. An airtight cabinet with sufficient space for all the specimens and glass containers of sulphuric acid are also required, together with vernier callipers accurate to 0.1 mm.

Procedure

The sample should first be conditioned to a given low relative humidity – say, 30 per cent, which would need sulphuric acid of relative density 1.43 in a small airtight cabinet (take care!). Equilibrium may take some days to reach and samples should be weighed periodically until their mass becomes constant. Mark pieces at a convenient position on the longest dimension and take calliper readings at this position. Readings should be taken several times to ensure repeatability and the mass of sample noted.

The samples are then returned and the humidity increased to say, 90 per cent, which would need sulphuric acid of relative density 1.14. Again weigh the samples periodically until the weight is constant. Soaking to 'speed up' the operation is not recommended as it leads to moisture gradients and may damage the sample.

When constant weight is achieved, the calliper readings and masses are again taken. Find the difference between the readings corresponding to the two humidities and represent the movement as a percentage. (Note that the difference in readings in the longitudinal direction will be very small.)

Finally, oven dry the timber at 105 °C and find the dry mass. Hence find the moisture contents of the timber at the two humidities. Compile a table giving radial, tangential and longitudinal moisture movements between the moisture contents concerned for each species.

Experiment 8.3   Visual stress grading of timber (BS 4978)

The basic technique of visual stress grading is very simple, though the process may be found to be time-consuming until experience is gained as to the critical criteria for any one piece. It is suggested that at least three pieces be obtained of nominal size 100×50 mm and length 2.0 m – this minimum length being necessary for distortion measurements. Pieces should ideally be selected so that at least one sample of each grade SS, GS and Reject is included. A simple scribe for measuring the slope of the grain is required – this comprises a cranked rod with a swivel handle and an inclined needle at its end (Fig. 8.25). A 3 m straight edge is also required.

Fig. 8.25   *Scribe with swivel handle for measuring slope of grain.*

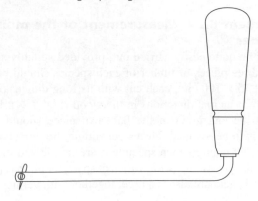

Procedure

Knots

Find the worst knot or combination of knots in the piece and mark the positions of surface intercepts on a sketch of the timber section, drawn full size on graph paper. Join the intercepts, inspecting the grain at the end of the piece to ascertain radial directions, if necessary. Hence obtain the knot area ratio (TKAR). Note also whether a margin condition is present (MKAR $> \frac{1}{2}$, Fig. 8.14).

Fissures

Inspect for fissures and record:
(a) position;
(b) size (depth) as a fraction of the thickness of the piece;
(c) length

Slope of grain

This should be measured on one face and edge. Press the needle of the scribe into the wood and draw along in the apparent direction of the grain. The needle will form a groove and give the precise direction. Avoid excess pressure which would prevent steady movement, or inadequate pressure which would cause the needle to 'jump' across the fibres. Note that growth ring direction and grain direction do not normally coincide. Measure the slope of the grain as in Fig. 8.26. The direction of fissures, when present, also indicate the grain direction.

Fig. 8.26 *Method of measurement of slope of grain.*

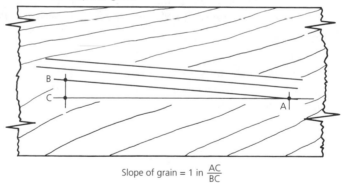

Slope of grain = 1 in $\dfrac{AC}{BC}$

Wane

The wane on any face is equal to the amount of that dimension that is missing. Express this as a fraction of the dimensions concerned. For example, if wane reduces the thickness of a 100 × 50 mm piece from 50 to 40 mm, the wane is 10 mm or one-fifth of the dimension.

Rate of growth

Measure the average width of growth rings at right-angles to growth, measuring rings over a 75 mm length if possible and omitting material within 25 mm of the pith.

Distortion

Cupping will not normally be a problem in pieces of section 100×50 mm. Measure twisting by weighting one end of the piece flat on the floor, such that the other end of the timber lifts. Measure the difference in uplift between the two edges of the piece (both ends will lift if the timber is bowed). Bowing and springing are easily measured using a 2 m straight edge (Fig. 8.8).

Assessment of grade

Compare results in respect of knots, fissures, slope of grain, wane, rate of growth and distortion with the requirements of Table 8.6. To comply with a given grade, pieces must satisfy all requirements. Any piece failing a GS requirement is designated 'Reject'.

Experiment 8.4 Machine stress grading

This experiment is not an attempt to simulate commercial machine grading tests in which performance of each localised section of the piece is assessed. It should be possible, however, to correlate results of this simple test with visual grading results, especially where the latter grades are determined on the basis of rate of growth or knot characteristics. For this reason it is recommended that the experiment be carried out on pieces of timber which have already been visually graded.

It is, of course, unnecessary to test timber to destruction – deflection tests are adequate, though destructive tests will yield bending stress results which can be compared with the stresses given in Table 8.8.

Pieces should be loaded on edge; for 3 m pieces of 100×50 mm softwood, this will require a machine with a load capacity in the region of 10 kN. A compression machine with a suitable steel beam insert could be used if a flexural machine is not available.

Procedure

Locate the piece centrally in the machine (Fig. 8.27) using $50 \times 50 \times 6$ mm steel bearers where timber touches the loading and support rollers. Fix a dial gauge under the piece at centre span.

Fig. 8.27 *Loading arrangement for machine stress grading.*

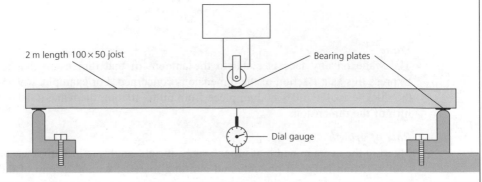

Apply the load progressively and record the load for each millimetre increase of deflection up to 10 mm. Remove the dial gauge and, if required, load the piece to destruction. Note the mode of failure. Repeat for the other pieces.

Results

Plot graphs of load (Y axis) against deflection (X axis) and draw the best straight line through the points. Obtain the gradient of the graph in N/mm. Note that higher gradients signify stiffer material.

The elastic modulus (E) of the timber may be calculated:

$$E = \frac{WL^3}{48\partial I} \qquad \text{where} \quad \partial = \text{deflection}$$
$$W, L, I \text{ as used earlier}$$

$$= \frac{W}{\partial} \times \frac{L^3}{48I}$$

$$= \text{gradient of graph} \times \frac{L^3}{48I}$$

The stress failure of the timber is

$$f = \frac{M}{Z} = \frac{WL}{8Z}$$

Calculate the stress at failure. Look for correlations between the elastic modulus and strength values for each piece and the visual stress grade result.

SELF-ASSESSMENT QUESTIONS

1. Describe the nature and function of the following:
   (a) vessels (hardwood);
   (b) fibres (hardwood);
   (c) tracheids (softwood);
   (d) rays.

2. (a) Describe the visual differences between hardwood and softwood by inspection of Figs 8.3 and 8.4.
   (b) Distinguish between ring-porous and diffuse-porous hardwoods and give one common example of each.

3. Comment on the strength, durability and availability of softwoods compared with hardwoods. Hence give main applications of each group in construction.

4. Explain the meaning of the term 'seasoning'. Give three advantages of seasoned timber over wet or 'green' timber. Explain why timber for internal use should be dried to a lower moisture content than roofing timber.

5. Describe the basic causes of distortion in timber. Suggest steps by which this distortion can be minimised in practice.

6. Describe three types of fungus attack in timber, giving situations where they commonly occur and their visible effects on timber. Outline methods by which the risk of decay can be minimised.

7. Describe what action should be taken when dry rot is found in the joists supporting a suspended ground floor.

8. Define the terms 'knot area ratio' (TKAR) and 'margin condition' (MKAR). Give the BS 4978 requirements for KAR values for 'General Structure' and 'Special Structural' grades.

Fig. 8.28 *Knot area ratio (KAR) classification.*

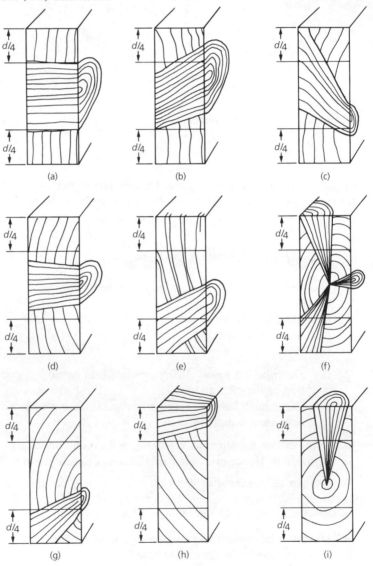

Estimate the KAR values of the sections shown in Fig. 8.28 and hence classify them in this respect.

9. Explain the principle of machine stress grading. Give two advantages of machine stress grading over visual stress grading.

10. (a) A uniformly distributed load of total value 25 kN is to be supported by 10 timber joists spanning 3 m. A joist size of 175 × 50 mm has been suggested. If the permissible bending stress in the timber is 5.3 N/mm^2, calculate whether this size of joist would be satisfactory from a bending stress standpoint. Ignore the self-weight of the joists

    (b) If the elastic modulus of the timber is 8600 N/mm^2, calculate the maximum deflection and check whether it is within the permissible limit of 0.003 × span.

11. A loft conversion is proposed which will result in loading on a stud partition at first floor level. The partition is of height 3 m and comprises 100 × 45 mm sections at 400 mm centres with bracing at mid-height by timber of the same section. A load of 5 kN per metre run is anticipated. Calculate whether the partition would be of sufficient strength if grade C14 timber is used (see Table 8.8).

12. Compare the properties of particle (chip)board and oriented strand board (OSB) for:
    (a) flooring;
    (b) roof decking applications.
    Give, in each case, a suitable BS EN specification code if it is anticipated that high relative humidity levels may be encountered.

13. Outline the main attractions of the use of glued laminated timber (glulam) for single-storey applications. State possible problems relating to multi-storey applications.

14. Discuss the options for timber from demolition sites to be reused or recycled.

REFERENCES

Building Regulations (1991) HMSO
National House Building Council Manual
Building Research Establishment Digest 393: *Specifying preservative treatments: a new European approach*, 1994

STANDARDS

BS 144: 1997: *Coal tar creosote for wood preservation*
BS 476 Part 7: 1997: *Method of test to determine the classification of the surface spread of flame of products*
BS 1282: 1975 (1997): *Guide to the choice, use and application of wood preservatives*
BS 4072: *Wood preservation by means of water-borne copper/chrome/arsenic compositions*
    Part 1: 1987: *Specification for preservatives*
    Part 2: 1987: *Method of timber treatment*

BS 4169: 1988: *Manufacture of glued laminated structural timber members*
BS 4978: 1996: *Visual strength grading of softwood*
BS 5268: *Structural use of timber*
    Part 2: 1996: *Code of practice for permissible stress design, materials and workmanship*
    Part 4: *Fire resistance of timber structures*
    Part 5: 1989 (1997): *Code of practice for the preservative treatment of structural timber*
BS 5589: 1989 (1997): *Code of practice for preservation of timber*
BS 5707: 1997: *Preparations of wood preservatives in organic solvents*
BS EN 300: 1997: *Oriented strand boards (OSB). Definitions, classifications and specifications*
BS EN 301: 1992: *Adhesives, phenolic and aminoplastic, for loadbearing timber structures; classfication and performance requirements*
BS EN 312: *Particle boards. Specifications*
BS EN 316:1993: *Wood fibre boards. Definition, classification and symbols*
BS EN 335: *Hazard classes of wood and wood based products against biological attack*
    BS EN 335–1: 1992: *Classification of hazard classes*
    BS EN 350–2: 1994: *Guide to the natural durability and treatability of selected wood species of importance in Europe*
BS EN 351–1: 1996: *Classification of preservative penetration and retention*
BS EN 460: 1994: *Durability of wood and wood based products. Natural durabilty of solid wood. Guide to the durability requirements for wood to be used in hazard classes*
BS EN 519: 1995: *Structural timber. Grading. Requirements for machine graded timber and grading machines*
BS EN 599: *Durability of wood and wood based products. Performance of preservatives as determined by biological tests.*
BS EN 588–1: 1997: *Specification according to hazard class*
BS EN 622: *Fibreboards. Specifications*

9 Plastics

Chapter summary

- Importance of carbon.
- Polymerisation of plastics.
- Classification.
- Thermoplastics, elastomers and thermosets.
- Plastics in fire.
- Applications.
- Recycling of plastics.

9.1 Introduction

The term 'plastic' refers to the fact that many plastics become soft and malleable when heated, though a more accurate description would be 'polymeric materials' which are almost always of organic origin; that is, based on carbon. These materials have in general the following attractions:

- They have low density – usually around that of water.
- An almost infinite range of plastics can be produced, based on the bonding versatility of the carbon atom.
- Many plastics are quite inert; they do not absorb water and are unaffected by frost, pollutants, dilute acids and alkalis. Hence surface protection from moisture is unnecessary.
- They are readily moulded or shaped.
- They are good thermal insulators, especially in foamed form.

Some shortcomings of organic polymers are:

- They are often susceptible to ultraviolet deterioration.
- They have high thermal movement.
- They have low stiffness (elastic modulus).
- They have generally low resistance to heat and some are susceptible to fire, especially in foam or sheet form.
- They rely (currently) largely on fossil fuels as the raw material.

Table 9.1 *Environmental considerations: plastics*

Consideration	Assessment
Raw material availability	Limited since plastics are in general a by-product of oil
Extraction	Risk of pollution – for example, from oil waste or discharge from sinking of ships
Energy used in manufacture	High, since chemical processes required
Health/safety hazards	Possible fire hazard; some plastics contain halogens
Waste disposal/recyclability	Many types do not degrade on disposal; recycling techniques are currently being developed

The technology of plastics is being continually developed and the solution to some of the above problems, at least in part, is likely in the foreseeable future. Environmental considerations are given in Table 9.1.

An understanding of plastics will depend very much upon a study of their chemical make-up and so a short section is devoted first to this.

9.2 Importance of carbon

The very important role of carbon in producing polymeric materials can be seen by reference to the simplified periodic table shown in Fig. 2.6 (page 27).

Carbon is the lightest element with four electrons in its outer shell which gives it great bonding versatility. It usually shares electrons with those of other atoms in order to produce a total of eight electrons (covalent bonding). This is easiest with atoms of similar size, for example, H, N, O, F, Cl to give a vast range of compounds normally described as covalent – the result of electron sharing. Silicon (element 16) likewise has four electrons but bonds less readily with lighter atoms because of its larger size. Hence the great majority of plastics are based on carbon.

To produce a solid material, bonding must be extended between thousands of atoms. Such bonding may result in regular patterns of atoms (crystals), irregular arrangements (amorphous materials) or long chains, which lead to formation of fibres.

Compounds based on carbon

As indicated above, the carbon atom (atomic number 6) is 4 electrons short of a complete shell and generally seeks therefore to share 4 electrons from other atoms. Bonding with hydrogen to produce hydrocarbons is particularly easy since the hydrogen atoms fit nicely around the carbon atom. The simplest hydrocarbon compound is methane, where one carbon atom is combined with four hydrogen atoms, as shown in Fig. 2.8. Another possibility is for two or more atoms of carbon to combine with hydrogen to form ethane, propane, or butane (Fig. 9.1).

Fig. 9.1 *Covalent bonding between carbon and hydrogen, forming ethane, propane and butane.*

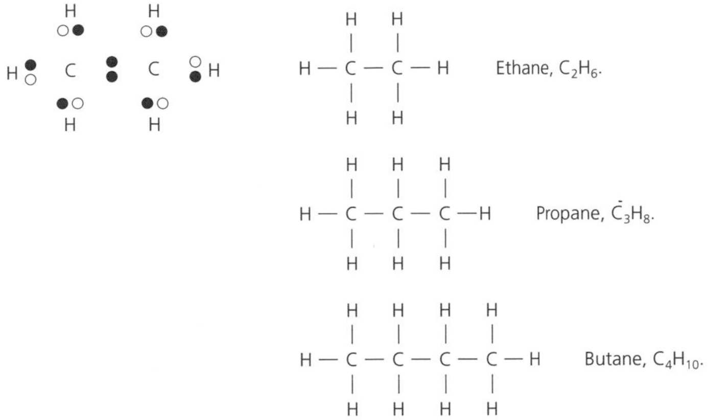

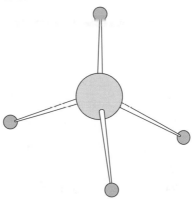

Ethane, C_2H_6.

Propane, C_3H_8.

Butane, C_4H_{10}.

Fig. 9.2 *Tetrahedral bond directions in carbon.*

This can obviously be extended indefinitely to form very-long-chain molecules, which become heavier as the size increases. It can be seen that the carbon atoms are joined by 'single bonds' corresponding to a single electron shared. Such compounds, where each carbon atom is linked to other atoms by single bonds, are known as **saturated compounds**. Each of the four carbon bonds is oriented in a particular direction, as shown in Fig. 9.2. The saturated compounds described are known as **alkanes** or **paraffins**, the four bonding arms trying to form a pyramid shape with equal angles between them. They have the general formula C_nH_{2n+2}. Another way of combining the carbon atoms is in the form of a ring, for example, benzene (Fig. 9.3). Benzene contains three sets of 'pairs' of lines. These correspond to **two** electrons being shared between the atoms concerned. They are known as double bonds and are not as strong as those represented by single lines (single bonds). However, they may be formed in certain conditions – for example, if the ratio of hydrogen to carbon atoms is too small for single bonds to form. Hence, benzene has the formula C_6H_6, requiring less hydrogen than the paraffins above. Examples of hydrocarbon bonding possibilities can be seen in Fig. 9.4. In each case it will be noticed that one of the 'arms' (bonds) is unconnected and this is

Fig. 9.3 *Benzene, C_6H_6.*

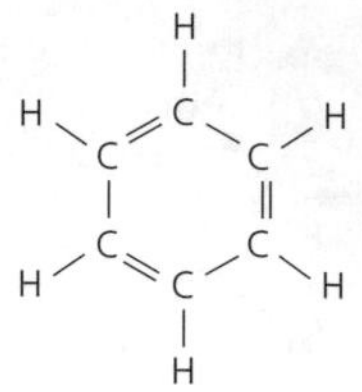

Fig. 9.4 *Hydrocarbon radicals.*

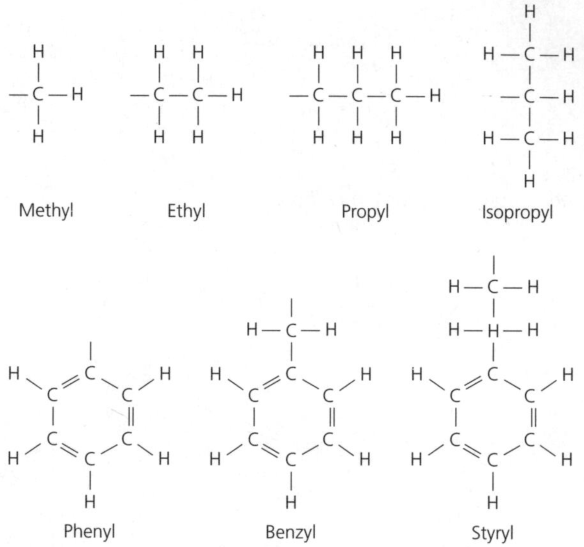

Methyl   Ethyl   Propyl   Isopropyl

Phenyl   Benzyl   Styryl

Fig. 9.5 *Radicals that combine with the hydrocarbons.*

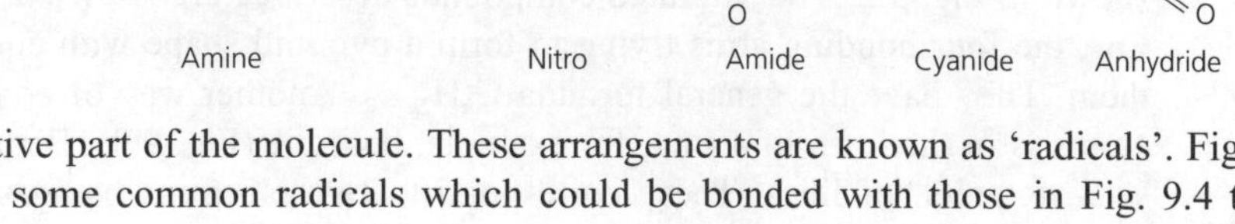

Alcohol   Aldehyde   Ketone   Ether   Acid
(phenol in aromatic compounds)

Amine   Nitro   Amide   Cyanide   Anhydride

the active part of the molecule. These arrangements are known as 'radicals'. Figure 9.5 shows some common radicals which could be bonded with those in Fig. 9.4 to form polymers.

Unsaturated compounds

As indicated above, it is possible for two or even three electrons to be shared between two atoms, forming double and triple bonds respectively. If an organic compound has

Fig. 9.6 *Ethylene (ethene), C_2H_4.*

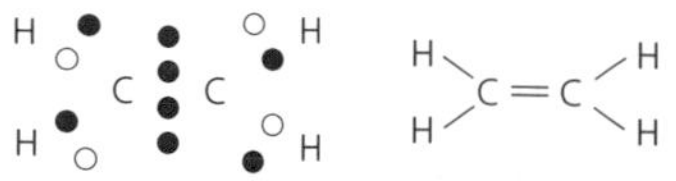

Fig. 9.7 *Acetylene (ethyne), C_2H_2.*

$$H-C\equiv C-H$$

one or more of its carbon atoms linked chemically in this way, then it is said to be **unsaturated**. Ethylene (ethene) is one of the simplest of the saturated organic compounds, as shown in Fig. 9.6. Because the four bonding 'arms' of the carbon atom are not spaced in three dimensions evenly around it, this involves considerable stress and is not the most stable arrangement; it would only happen if there were a shortage of H atoms at the time of chemical combination. An example of a triple bond is shown in Fig. 9.7 – acetylene (ethyne). These multiple bonds in unsaturated compounds are easily broken leaving vacant 'arms' ready to combine with other groups or elements. This fact is utilised in the production of some very-long-chain polymers.

9.3 Polymerisation

In order to produce the solid state very large molecules need to be produced.

Polymerisation is the linking together of simple molecules to form larger ones. The original molecular unit is described as the **monomer** and the resulting molecule is known as a **polymer**. The process is accompanied by a gradual change from the liquid state to the solid state.

Addition polymerisation

This process occurs when molecules of a monomer such as ethylene react additively to produce long-chain molecules with possibly as many as 20,000 links in the chain. This can be achieved by means of initiator molecules which effectively contain highly reactive 'free radicals':

$$I-I \longrightarrow I^- + I^-$$

Initiator mol.   Free radicals

The introduction of a small amount of initiator into ethylene (Fig. 9.6) opens the double bonds:

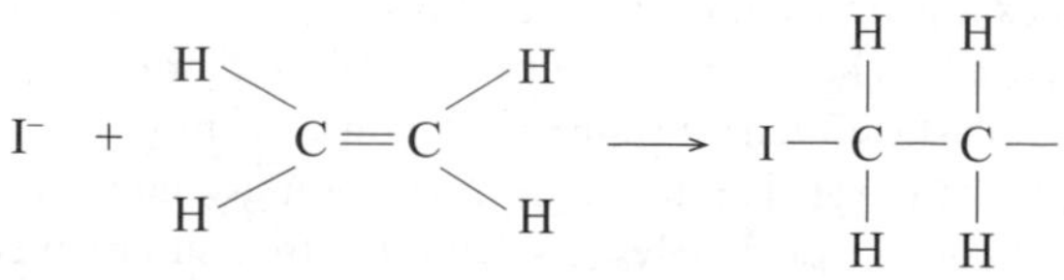

This highly reactive compound can now react with another ethylene molecule:

$$
\begin{array}{ccc}
\underset{\displaystyle \overset{\displaystyle}{}}{\text{I}-\text{C}-\text{C}-} & + & \text{C}=\text{C} \\
\end{array}
\longrightarrow \ \text{I}-\text{C}-\text{C}-\text{C}-\text{C}-
$$

The process continues rapidly with molecules continuing to grow until all the ethylene is polymerised, polymer chains terminating in initiator atoms.

Condensation polymerisation

This is a polymerisation that usually involves two different types of monomer. The molecules of at least one of these monomers contain two or more reactive groups of atoms or radicals, while the molecules of the other may contain one or more reactive groups. Chemical interaction of these monomers produces plastics materials which have either long-chain molecules or a three-dimensional cross-linked structure, depending on the number of reactive groups. Larger numbers of reactive groups are more likely to produce three-dimensional molecules. In either case, molecules of such simple substances as water (H_2O) or hydrogen chloride (HCl) are eliminated. It is in this respect that condensation polymerisation differs from addition polymerisation. Figure 9.8 shows the polymerisation of phenol and formaldehyde producing 'Bakelite' – one of the earliest plastics to be produced. Each phenol molecule has three reaction sites – alternate carbon atoms. Formaldehyde has two reaction sites corresponding to the double bond with the oxygen. When any two phenol molecules and one formaldehyde molecule come together, a water molecule is released. A rigid three-dimensional molecule is readily built.

9.4 Classification of plastics

There are two groups of plastic: thermoplastics and thermosetting plastics (thermosets).

Thermoplastics

Thermoplastics can be softened by heating after they have been cured, and can be remoulded, if desired.

For manufacturing purposes this has the big advantage that the polymer can be produced in convenient bead form and then supped to the fabricator to manufacture the product required. In molecular terms thermoplastics comprise long-chain molecules, lightly bonded by forces arising from charge eccentricities on each polymer chain (van der Waals forces, Fig. 2.12). These forces are overcome by thermal energy on application of heat so that the plastic softens. Problems of bonds of this type are that strength and stiffness are in general lower than those of more strongly bonded materials, while thermal movement is higher – some polymers have coefficients of thermal expansion

Fig. 9.8 *Formation of phenol formaldehyde resin.*

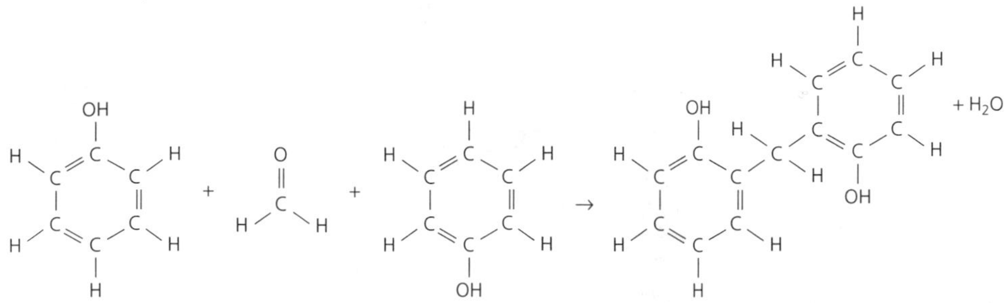

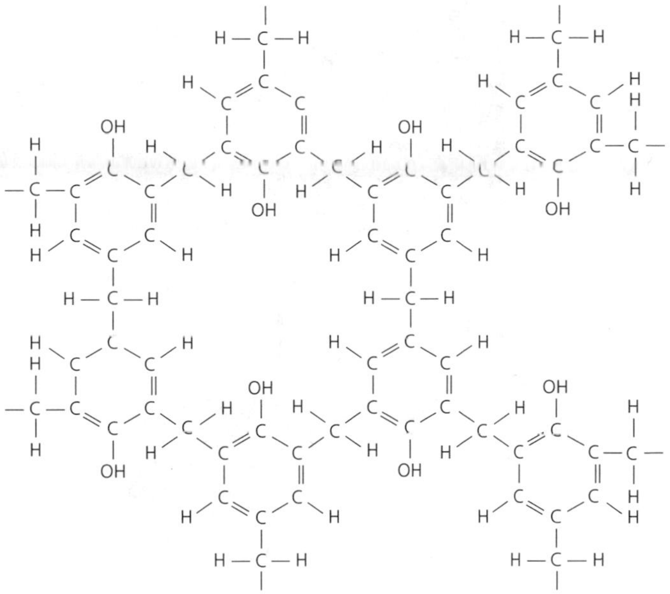

Phenol   Formaldehyde   Phenol   Preliminary condensation products

Phenol formaldehyde resin

over $100 \times 10^{-6}/°C$ (compared, for example, with $11 \times 10^{-6}/°C$ for steel). Nevertheless, thermoplastics are generally cheaper than thermosetting plastics and are more widely used in the construction industry. They can be formed by:

- **Injection moulding**, in which the softened polymer is injected under pressure into a heated mould. This process would be used to form articles of irregular shape such as gutter brackets.
- **Extrusion**, in which the softened material is forced through a die to give long lengths of constant section. Such sections can be quite complex, producing the maximum possible rigidity for a given polymer type (Fig. 9.9).

Fig. 9.9 *Typical window extrusion in uPVC. The joint is made by mitring two sections, heating and then pressing them together. The extrusions are of a complex shape to produce structural stability and to make provision for fixing of rubber gaskets, and steel reinforcement, should it be required.*

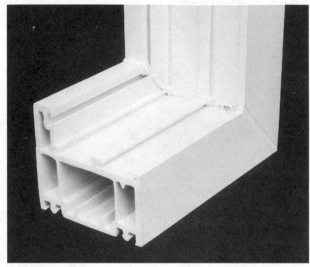

Fig. 9.10 *Growth of spherulites containing crystalline regions in polyethylene chains.*

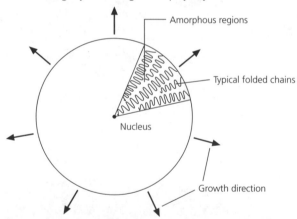

Common examples of thermoplastics are polyethylene (polythene), polypropylene, polyvinyl chloride, polymethyl methacrylate (Perspex), polystyrene and nylon. Two important concepts which should be appreciated in relation to thermoplastics are *crystallinity* and the *glass transition temperature*:

Crystallinity

In simple linear polymers, the molecular chains have a tendency to fold back on themselves as they form, chains interleaving and producing relatively dense packing which might be described as a crystal (Fig. 9.10). The result of this is increased bond strength and, hence, increased rigidity for a given polymer. Polymers such as polyethylene,

Fig. 9.11 *Alternating of bond directions in polyethylene chain so as to minimise repulsion between hydrogen atoms. Above T_g all C–C bonds can rotate as shown.*

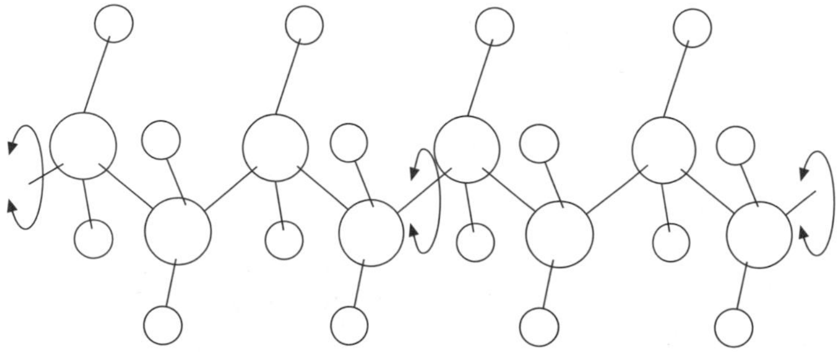

polypropylene, nylon and polytetrafluoroethylene (PTFE) crystallise to some extent, their mechanical properties improving in consequence. Since the crystallites tend to intercept light, these polymers can never be obtained glass-clear; even in thin sheet form they have a milky appearance. Polymers with a more random distribution of radicals, or more complex groupings on the carbon backbone, cannot crystallise. These (amorphous) polymers must rely on van der Waals bonds for rigidity, though, in consequence, they will be available in glass-clear sheet form (assuming that they do not contain fillers which would interfere with transparency). Most thermoplastics fall into this category.

Glass transition temperature

The extension or compression of polymers often results in a twisting action at molecular level since polymer chains are not straight – they may take up a combination of zig-zag, folded or helical configurations. Above a certain temperature, known as the glass transition temperature (T_g), the polymer atoms have sufficient thermal energy to enable the molecular chains to twist (Fig. 9.11). This involves hydrogen or other side chain groups passing each other, overcoming the repulsive forces which tend to keep them apart. In consequence the flexibility of the polymer increases, and it becomes much more pliable. Conversely, polymers at temperatures below their T_g tend to be stronger and stiffer, but more brittle. Thermoplastics for use in applications requiring some rigidity may be used in two chief modes:

- Amorphous polymers are generally employed below their T_g value (which is about the same as their softening point) since, when above this temperature, van der Waals forces are insufficient on their own to produce rigidity. Polymers such as rigid polyvinyl chloride, polystyrene, acrylic plastics, acrylonitrile butadiene styrene (ABS) are important examples of amorphous thermoplastics well below their T_g at room temperature.
- Crystalline polymers, with their extra rigidity due to their crystalline regions, can be used above their T_g. Polymers used in this state usually have considerable flexibility due to permitted bond rotation. While not, in general, being suitable for structural applications, they may be used where flexibility and impact resistance are required. Common examples include polyethylene (polythene), polypropylene and polytetrafluoroethylene (PTFE).

Table 9.2 gives properties of some common thermoplastics.

Table 9.2 *Common thermoplastics: properties*

In order of decreasing elastic modulus

Polymer	Degree of crystallisation (%)	T_g (°C)	Relative density	Expansion coefficient (10^{-6} °C)	Short-term tensile strength (N/mm²)	Tensile elastic modulus (kN/mm²)
Polyvinyl chloride	0	80	1.40	70	50	3.2
Polystyrene	0	80–100	1.40	70	50	3.0
Polymethyl methacrylate	0	80–100	1.18	65	70	2.9
Polycarbonate	0	150	1.22	55	60	2.7
Nylon 66	100	57	1.14	90	70	2.6
ABS	0	70–110	1.05	80	40	2.5
Polypropylene	60	−27; 10	0.90	110	30	1.3
HD polyethylene	95	−120	0.95	130	27	0.9
PTFE	95	−120	2.10	110	20	0.5
LD polyethylene	60	−120	0.92	220	8	0.5

Thermosetting plastics

Thermosets cannot be softened by heat; hence the products must be moulded prior to polymerisation.

This is usually achieved by mixing a resin and hardener and moulding fairly quickly. After hardening the moulded shape cannot be deformed by heating – the material begins to break down chemically before it softens. Thermosets comprise giant three-dimensional molecules (Fig. 9.8) leading to higher strength and stiffness than thermoplastics, together with better resistance to organic solvents. However, this rigidity often leads to brittleness and unless thermosets are toughened – for example, with fibres – situations in which impact is a possibility should be avoided. Examples are phenol formaldehyde, urea formaldehyde, polyester resins and epoxy resins.

There is in fact no rigid distinction between thermoplastics and thermosets, some polymers being lightly cross-linked to form materials with intermediate properties. Polyesters, for example, can be obtained in a flexible fibre form as used in fabrics or in the much more rigid form as the matrix of GRP – glass-reinforced polyester. Rubbers, another name for elastomers, which comprise zig-zag or helical molecules, can be stiffened by vulcanising, in which chains are cross-linked by means of sulphur atoms.

9.5 Some common thermoplastics

Basic composition, properties and uses of the common thermoplastics are given below. Properties may be compared by reference to Table 9.2.

Polyethylene – polythene $(CH_2-CH_2)_n$

This may be regarded as the simplest thermoplastic, being a plain hydrocarbon chain of great length. The structure is highly covalent and, in consequence, van der Waals forces between the chains are small since there are few of the charge eccentricities which are required for them to operate. The regular chains have, however, a tendency to crystallise, folding back on each other during formation and packing closely together. This produces sufficient cohesion to form a solid, though with a low stiffness and melting point (polythene is above its T_g at room temperature). Two types are available – **high-density** polythene is more crystalline and therefore stronger and stiffer than the **low-density** material, which is less crystalline. However, since both are pure hydrocarbons, each floats on water; that is, they have densities less than 1000 kg/m³. Polythenes have a waxy feel, indicative of the covalent nature of their bonding, and cannot be joined by adhesives due to their low surface chemical activity. They are embrittled by ultraviolet light and so must be protected, which is usually achieved by incorporation of carbon black which absorbs the UV rays, avoiding damage. The most important application of low-density polythene is in sheeting for resisting moisture penetration. The material is used both in damp-proof courses and damp-proof membranes. Its toughness in these situations is an advantage, combined with the fact that UV rays should not be a problem. High-density varieties can be used in underground pipes, where their chemical stability and toughness are attractions, though substantial wall thicknesses are essential in view of their low-tensile strength.

Polypropylene $(-CH_2-CH\cdot CH_3-)_n$

This is similar to polythene except that one H atom is replaced by a methyl group. This leads to higher chain interaction which, together with crystallisation, produces a stiffer, stronger polymer with a higher melting point. Like polythene, it also floats on water, being a pure hydrocarbon. Also, like polythene, it has good chemical resistance and is used for waste systems, including laboratory wastes, and moulded articles such as manholes. It cannot be solvent welded. A further important application is as a fibre used to toughen concrete; the drawn fibre has a tensile strength approaching that of mild steel (though the stiffness is much lower) and the fibres appear to act as crack arrestors in brittle materials such as concrete.

Polystyrene

This differs from polythene in that it contains benzene rings (Fig. 9.3) which greatly increase chain interlock and rigidity, though a slightly brittle material (amorphous, below T_g) results. Its most important form is as a rigid or granular insulation material, where its rigidity, even in expanded form, enables it to withstand handling stresses.

Normal grades are highly flammable and constitute a serious hazard when being stored on construction sites or when fixed in living areas in the form of expanded polystyrene tiles – especially on ceilings. Flame-retardant grades are available. Earlier versions incorporated halogens which achieve this function but they may release toxins in the event of fire and also increase the hazards of disposal. Halogen-free retardants are now available. In tiling applications fire-retardant types should be used and tiles should always be fixed with a continuous film of adhesive rather than dabs, which, in the case of fire, would permit the tile to drop easily and 'feed' the fire.

A much less hazardous application is in the form of expanded sheeting fixed by adhesive *behind* plasterboard, where the latter affords adequate fire protection. Rigid polystyrene 'bats' are widely used as cavity insulation, permitting a cavity to be retained. It can be used in bead form for retro cavity filling, the beads flowing readily and therefore requiring fewer holes to be drilled than for other fill materials. However, the plasticiser from any plasticised PVC electricity cables in a cavity can leach into the polystyrene resulting in embrittlement of the cable.

This type of insulation is therefore not recommended where electrical cables are run through the cavity.

Polyvinyl chloride (PVC) (–CH$_2$–CH·Cl–)$_n$

Chlorine, with its one electron vacancy, is much more likely to produce ionic bonds. The chain, therefore, becomes electrically charged leading to stronger inter-chain interaction. It also reduces the flammability of the product. Consequently, PVC (amorphous, below T_g), with its greater strength and rigidity and reasonable cost, is the most widely used thermoplastic in the construction industry. It has the further advantage of being amenable to solvent welding as well as joining by thermal fusion (Fig. 9.9).

A major development in the last 20 years has been uPVC windows. Formulations of the materials tend to be evolutionary, involving fillers, additives to control UV degradation and impact properties, and pigments to obtain the desired colour. Complex extrusions forming the basis of such frames can be cheaply produced. They can incorporate slots for glazing seals and for steel reinforcement strips if required. Larger frames, or uPVC doors, should contain such steel reinforcement to ensure adequate strength and resistance to distortion by creep. Corner jointing is straightforward – ends are simply mitred and heated by contact with a hot metal plate. An accurate thermally fused joint is then obtained. Glazing should be sealed by rubber gaskets which form a good seal between the smooth uPVC/glass surfaces and permit the relative thermal movement that can be expected between them. The good-quality maintenance-free surface finish is an attraction and this, combined with their competitive cost, has resulted in very widespread use. Life is limited to around 20 years depending on conditions. Darker frames probably have reduced life because they get hotter in warm weather. Excessive heat in severe summer conditions can cause distortion since thermoplastics have a 'memory' and tend in such conditions to revert to their former molecular arrangement prior to extrusion. Further extensive uses are in waste systems, where solvent welding provides a quick method of jointing, and rain-water systems, where installations can be provided at a fraction of the cost of other systems.

Although most commonly used in unplasticised form (uPVC), organic plasticisers can be introduced to produce softer sheets or fabrics. (These in effect lower the Tg to below room temperature so that a much more flexible product is obtained.) In common with other thermoplastics, PVC is degraded by ultraviolet light, though stabilisers can be included.

Polytetrafluoroethylene (PTFE) $(-C_2F_4-)_n$

Substitution of all the H's in polyethylene for fluorine produces a stronger chain attraction which, together with crystallisation, results in a polymer with great chemical stability, low mechanical friction, non-flammability and quite a high melting point. PTFE is most commonly used as a pipe-jointing tape where it acts as a thread sealant and assists in tightening the joint. More recently is has been used, reinforced with fibreglass fabric, in fabric roofs (Fig. 2.16).

Nylons (polyamides)

These contain amide groups (see Fig. 9.5) resulting in strong chain attraction, which, together with crystallisation, produces a range of high-performance polymers with melting points over 200 °C. Applications in construction are restricted on account of cost, but include fasteners and tubing.

Polymethyl methacrylate (PMMA) (Perspex)

This is based on acrylic acid (CH_2–CH.COOH) which polymerises to form strongly attracted chains, resulting in a hard, rigid, amorphous polymer. The plastic has glass-like transparency, can be finished to a high lustre and has a relatively hard surface and good weathering resistance. It is used for baths, glazing and shop signs. It is also used in paints, as a binder in resin concrete and for crack repair. Both the resin and the polymer are highly flammable so precautions must be taken where appropriate.

Acrylonitrile butadiene styrene (ABS)

This is an example of a copolymer combining the rigidity of styrene with the toughness of butadiene styrene rubber. The material is rather similar to uPVC, though it is tougher, more heat resistant and more expensive. It can be solvent welded with an appropriate solvent adhesive. Examples of use include waste fittings, where it is able to withstand higher effluent temperatures than uPVC.

Others

Examples of further plastics developed for specific applications include:

- Polyvinyl fluoride (PVF) – for durable plastic coatings.
- Polycarbonates – tough and rigid glazing.
- Modified PVC products for hot water pipes.

9.6 Elastomers

Elastomers are polymers such as rubbers with very high strain capability often well over 100 per cent.

Elastomers are generally thermoplastics above their T_g, which permits bond rotation. Molecules usually form zig-zag configurations so that stretching an arrangement of the type shown in Fig. 9.12 produces increasing resistance initially, followed by much greater resistance as chains approach a straight profile. This behaviour is typical of rubbers which, initially, have very low E values of between 1 and 20 N/mm^2 at strains up to 400 per cent. On further stretching, the E value rises as resistance is encountered

Fig. 9.12 *Arrangement of polymer chains: (a) random zig-zag profile of carbon chains in unstressed elastomer; (b) chain straightening caused by stretching.*

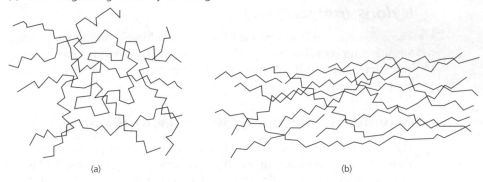

(a) (b)

Fig. 9.13 *Sealants can only accommodate large movements if they are permitted to change shape. The 'ideal' shape for most purposes is when breadth b is half depth d.*

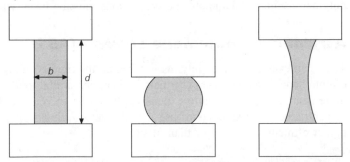

in the straightened chains, prior to failure. This effect is easily observed in a rubber band or balloon. Stretching beyond a certain point results in much more resistance to further stretching.

The low E value of elastomers is very important; higher values would generate stress both in the elastomer and in the materials with which they are used in conjunction.

At very high strain levels elastomers change shape; therefore, provision must be made for this to occur in use.

This means that when used, for example, as sealants, bonding on all sides should be avoided so that they can be allowed to change shape as movement is encountered (Figs 9.13 and 9.14). Some common elastomers are given below.

Fig. 9.14 *Incorrect and correct use of elastomers for sealing a crack/joint: (a) will fail because the sealant cannot change shape; (b) the joint is opened out and a debonding strip is used to prevent bonding behind the sealant.*

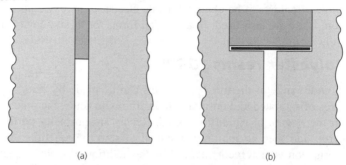

(a)   (b)

Natural rubber

This generates very little heat on flexing. It is used for bridge and machine bearings; even whole buildings have been mounted on rubber to isolate them from vibrations, for example, from underground trains.

Silicone rubbers

These are widely used as sealants. They can be obtained as two-part formulations, having the advantage of rapid curing, or as RTV (room temperature vulcanising) types which cure slowly by atmospheric moisture. Each is characterised by good adhesion, good elastic properties, water repellency and durability.

Polychloroprene (Neoprene)

This has good all-round chemical stability and, since it crystallises on stretching, it combines high extensibility with high strength. It has relatively good fire resistance. It is used in contact adhesives and as sealing strips for glazing and O rings.

9.7 Some common thermosets

Phenol formaldehyde

Phenolic resins were among the earliest to be produced; they were formerly made from coal tar and given the name Bakelite. After reaction with formaldehyde (HCOOH) the benzene ring in phenol provides three reactive sites at which polymerisation can occur by the condensation reaction shown in Fig. 9.8. A large three-dimensional molecule is produced, leading to a stiff, brittle polymer. All forms are dark brown or black in colour and are used where hardness and heat resistance are required; for example, in electrical goods such as switches and sockets. They are also used in laminates such as Formica, paper being used to toughen and reinforce the resin. A further application is in waterproof wood adhesives (for factory use since the hardener is highly acidic).

Urea formaldehyde

These resins are similar to phenolic resins except that they are clear, enabling a large range of colours to be obtained, white electrical goods being the most common application. They are also used, in expanded form, for in-situ cavity fills.

Polyester resins (GRP)

A wide range of thermosetting resins can be made by using styrene to cross-link mixtures of saturated and unsaturated dicarboxylic acids, the precise formulation depending on the degree of rigidity required. Resins are normally partly polymerised to form a syrup and then an inhibitor is added to stop the reaction. At the point of use, polymerisation is then recommenced by the addition of a *small* quantity of initiator such as organic peroxide. Since they are brittle on their own they are invariably reinforced with glass fibres which both toughen and stiffen the resin. Glass fibres have an E value of 70 kN/mm^2 compared with about 4 kN/mm^2 for the resin. The performance of the resulting composite, glass-reinforced polyester (or GRP), depends on the proportion of fibres; hand lay-up techniques, as used for car body repairs, contain only 10–20 per cent by volume of fibres, but by use of woven roving or winding of unidirectional fibres, E values of up to 40 kN/mm^2, with good tensile properties, can be obtained. Polyesters are subject to about 7 per cent shrinkage and this can affect adhesion to other materials. Surface glass fibres should be protected by a 'gel' coat to prevent admission of water, which causes deterioration of the fibres. The performance of GRP can only be as good as the protective gel coat, which must therefore be properly maintained in an exposed environment. GRP has been used for a wide range of precast moulded products, from inspection chambers to cladding panels and even precast portal frames for swimming pools.

Epoxy resins

These contain the epoxide ring:

$$R - CH - CH_2$$
with the O bridging across CH and CH₂

the ring being opened to form a polymer in much the same way as a polyester. The final polymerisation process is, however, different from polyesters in that the resin and hardener, having the action of 'hooks and eyes', become linked together in the resultant polymer (Fig. 9.15). It is important to have equal numbers of hooks and eyes and therefore important that precisely measured quantities of resin and hardener are mixed. This best achieved by the use of prebatched containers, carefully labelled, the total contents of the resin and hardener containers being used. Also, unlike polyester resins, most shrinkage is pre-gelation, post-gelation shrinkage only being about 1 per cent, so that good bonding properties are obtained. Epoxy resins have excellent resistance to chemicals, particularly to alkalis, and are used in adhesives ('Araldite'), heavy-duty flooring compositions, resin concretes and glass-reinforced composites. Their high cost is offset by very good performance.

Fig. 9.15 *Epoxy resin: (a) simple representation – epoxide groups being represented by 'eyes'; (b) hardener provides 'hooks' to link up with the 'eyes'. For good results the number of hooks and eyes must be equal.*

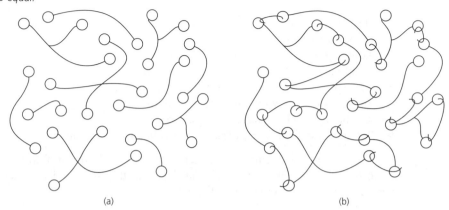

(a)   (b)

Polyurethanes

These are formed as a product of isocyanates and polyesters and polyhydroxy materials (see also Chapter 10, 'Paints'). It is not possible to classify polyurethanes as thermoplastic or thermosetting, since many different forms exist: some have chain-like molecules and are therefore, thermoplastics; others have heavily cross-linked molecules and are therefore thermosets. Polyurethanes are often used in foamed form, for example, in cavity filling for domestic dwellings and in a more flexible form for spraying on pipes for thermal-insulation purposes. In each case, they bond well to the background. Most polyurethane is foamed, using carbon dioxide, which eventually diffuses outwards from cells, being replaced by air. Polyurethanes may be used up to temperatures of about 120 °C depending on the grade – an advantage over polystyrene foams. They are also less flammable, though, in common with polystyrenes, more flexible types dissolve in some organic solvents and acids.

Flexible polyester and polyether foams come under the general heading of polyurethanes, being used for upholstery, sponges and cushions. Other polyurethanes are used for resilient gaskets in clay pipes and in surface coatings.

9.8 Plastics in fire

Since virtually all plastics are organic in nature, they decompose readily in fire and, consequently, many of them constitute a hazard, possibly for the following reasons:

1. They may emit toxic gases – usually carbon monoxide.
2. They often contribute to the development of fires by flaming and/or heat emission.
3. They may emit dense smoke, thereby making escape more difficult.
4. Melting of sheet glazing materials may vent the fire, increasing the rate of spread.

The fire hazard associated with plastics depends, in a complex way, on the type of plastic and its mode of use. It is found that many plastics perform badly in several of the above respects:

- Plastics which **burn readily** include polyethylene, polypropylene, polystyrene, acrylic resins, ABS, polyurethane, GRP and rubber.
- Acrylic polymers are an additional hazard because their **flash-ignition temperature** is only 290 °C; lower than that of any other common polymer.
- Some plastics which do not burn readily, nevertheless, produce **dense smoke** in fire; for example, polyvinyl chloride, which also produces hydrogen chloride gas. Polystyrene and GRP produce dense smoke.
- Polyurethane produces **hydrogen cyanide gas** on combustion.

The greatest fire hazard is, in general, caused by plastics in **foamed** or **sheet form**, since oxygen is much more readily available in this state and their temperature rises very quickly owing to their low thermal inertia.

- Serious disasters have been caused, for example, by combustion of vertical acrylic sheeting, due to the extremely rapid spread of flame in this situation.
- Many house fire fatalities have been caused by polyurethane foams in cushions, due to rapid combustion and the toxic fumes emitted. Standard foam may reach temperatures exceeding 700 °C within two minutes of ignition, with fatal levels of hydrogen cyanide emission. High-resilience foams containing more than two isocyanite groups are more heavily cross-linked and, therefore, decompose less easily but the combustion-modified high-resilience (CMHR) foams are much preferred. They may, for example, contain a melamine resin which acts as a flame retardant. Another type contains expandable graphite particles which quickly form a thick protective layer when foam is exposed to heat.
- Expanded polystyrene ceiling tiles are not recommended in areas where there is a fire risk, since they melt easily and may feed the fire, especially if the adhesive is used in 'dabs' rather than as a continuous film. Fire-retardant grades of these polymers, as well as GRP, polypropylene and polystyrene, are available but have not in the past been widely used on account of increased cost.
- Many thermoplastic adhesives or resins contain flammable solvents, constituting a considerable hazard during use and until solvent evaporation is complete or the resin is fully cured.
- The risk with some materials will depend entirely on the mode and situation of use. Plastics such as PVC, which do not burn well, may help to retard or accelerate the development of fire, for example:
  1. Collapse of PVC down pipes may prevent their transmitting fire vertically.
  2. Collapse of thermoplastic roof lights could reduce the build-up of heat in an enclosure, hence, delaying flash-over.
  3. An internal PVC soil pipe that is not adequately protected could melt and then act as a chimney, sucking additional air into a burning enclosure.

There are, however, many situations in buildings in which the use of plastics may be regarded as safe – for example, guttering, sheet roofing, floor coverings, decorative laminates, cavity insulation, water pipes, cisterns and underground pipes.

9.9 Application of polymers in the construction industry

Typical applications are given in Table 9.3 together with advantages and precautions applicable to each type.

Table 9.3 Common plastics application in the construction industry

Application	Polmer used	Attractions	Possible problem areas
Window frames	Unplasticised PVC	Attractive, maintenance-free surface; good thermal insulation; competitive cost; easily made to fit existing opening	Ensure adequate stiffness, especially if double glazing used; larger sizes should be steel reinforced to avoid creep; colour/finish deteriorates with time
Rain-water goods	Unplasticised PVC	Low cost; easy fixing; lightweight; range of designs/colours	Low rigidity compared with metal systems; movement joints must be allowed; adequate support needed; possibility of leaks/separation of sections after long service; age may cause embrittlement
Waste systems	Unplasticised PVC	Low cost; easily assembled using solvent cement	Limited resistance to hot effluent/solvents; long lengths may be subject to thermal movement/creep problems; joints cannot be opened after assembly
	ABS	Higher temperature/solvent/impact resistance	Higher cost; special solvent needed for jointing
Glazing	Acrylic plastics	Low cost; high impact resistance; low risk of injury on impact	Fire risk in some situations; not for use in large glazed areas; scratches easily
Cold-water cisterns	Glass-reinforced polyester	Light; durable; easily fixed	Good support necessary; avoid stressing at pipe connections
Cavity insulation	Polystyrene bats	Low cost; easily incorporated during construction; cavity can be retained	Some skill/supervision required to ensure correct installation

	Polmer used	Attractions	Possible problem areas
	Polystyrene beads	Can be incorporated into existing cavities	All openings (for example, for services) must be sealed prior to insertion; PVC-coated wiring in cavity may lose plasticiser to beads causing embrittlement; slight risk of damp penetration in severe weather
	Urea formaldehyde foam	Easily incorporated at low cost into existing cavities	Not recommended for severe exposure; foam may crack, leading to moisture admission; slight risk of formaldehyde gas in interior for a time after insertion
Foam filling for upholstery	Polyurethane foam	Low cost; good comfort level	Serious fire risk; unmodified foams no longer permitted; though much foam still in use. CMHR (combustion modified high resilience) foams now required by law
Fabrics for air-supported /tension structures	Polyester reinforced PVC	Relatively low cost; material melts in fire venting smoke	Life limited to 15–20 years
	Glass fibre-reinforced PTFE	Life up to 50 years; self-cleaning	Expensive; material remains intact in fire except in extreme heat; risk of build-up of toxic fumes

9.10 Recycling plastics

There are compelling reasons for recycling plastics wherever possible:

- Many types do not readily degrade in the ground and, as such, constitute an additional environmental hazard when used in land-fill.
- Organic polymers are a precious and finite resource.
- Many plastics in construction occur in forms in which separation from other waste is reasonably simple.

Nevertheless, at the present time, most recycling is restricted to household waste and there is a need to extend this to the refurbishment/demolition stages of construction.

One possible problem is the identification of the various common plastics types and this is now assisted by letter and numerical codes to be found on many plastics for ease of sorting. The codes are given in Table 9.4. Another mark found on some packaging is the 'Green Dot' logo (Fig. 9.16), which is a German scheme. Manufacturers license the use of the logo, and their license fees are used to fund the collection and sorting of packaging materials for recycling.

Plastics type	Number code	Letter code	Main applications
Polyethylene terephthalate*	1	PET	Bottles, food packaging, cling film
High-density polyethylene	2	HDPE	Tubing
PVC	3	V	Rain-water and waste systems
Low-density polyethylene	4	LDPE	Sheet materials
Polypropylene	5	PP	Tubing, plumbing fittings
Polystyrene	6	PS	Insulation, refrigerator linings, containers
Other types	7		

* The first type, polyethylene terephthalate, though not used in construction, is included for completeness and because it is an obvious candidate for recycling

Fig. 9.16 *The 'Green Dot' logo. (Reproduced courtesy of Duales System Deutschland AG)*

Since oil is the raw material for most plastics, and the material itself accounts for a substantial proportion of total cost, the imperative for recycling will depend upon the world economic climate and, in particular, oil prices. These have been depressed for a number of years and at present the cost of recycling poor-grade plastics waste may be higher than that of producing the original material. However, if prices rise, the attractiveness of recycling will greatly increase and as some plastics, such as PVC, have been in use in the construction industry for over 30 years, amounts of waste from this source are beginning to rise. In view of the fact that the quality of recycled material is likely to be, in general, lower than that of 'virgin' material, an intermediate goal would be to incorporate a percentage of recyclate in new products. An alternative would be to produce, for example, PVC window sections with a recycled core only, surface layers being virgin material. As sorting and recovery processes improve, this percentage could be increased.

The complexity of operations to recycle materials such as plastics will be illustrated by considering possible stages in the process.

Sorting

This separates plastics from other types of waste and, if possible, into different plastics types. Sorting techniques may be automated – for example, the plastic could be identified by spectroscopic or X-ray techniques and sorted accordingly, though articles which contain more than one type would be a problem. In addition, it may be necessary to remove other undesirable constituents, such as inks, prior to further processing.

Unless sophisticated sorting techniques are used there will still be a 'mix' of plastics types, so it may not be possible to recreate the original type of product made.

The sorted plastics may be dealt with in one of the three methods discussed below.

Mechanical recycling

In this process the plastics are subject to physical changes only. Since many plastics types are incompatible with each other, the process is currently limited to articles such as bottles, fully sorted and of known composition. (Plastic bottles form a major part of the world market for plastics.) The material would be pelletised and then softened to form new products. This is sometimes called 'primary' recycling.

Although there are limitations to the miscibility of plastics, mixtures within certain compositional limits can also be combined thermally to form a 'melt'. Products of such processes do not have the quality of pure plastics but could be reused for less demanding applications (e.g. wood substitutes). This is sometimes called 'secondary' recycling.

Stabiliser levels may need to be restored for outdoor applications since these are progressively lost during service.

It should be at least theoretically possible to recycle most thermoplastics by mechanical recycling since they all soften on heating. Such processes, where feasible, are preferred since they require much less energy than chemical reprocessing.

'Feedstock' recycling

The material is reprocessed into 'feedstock' for re-manufacture or some other process. Unsuitable material must be removed prior to chemical processing. Possibilities include:

- **Pyrolysis** – heating in the absence of air which causes polymer breakdown.
- **Hydrogenation** – heating under pressure with hydrogen which generates saturated hydrocarbons.
- **Gasification** – heating in a limited supply of air which produces mainly hydrogen gas. There is also a heat energy by-product.

These can be used for further plastics production or alternative purposes. Since chemical techniques are involved, the processing energy, combined with costs of initial sorting and cleaning, may not always be economically or environmentally viable. Thermosets are not softenable on heating and so would need to be treated in these types of process to permit recycling. It is sometimes called 'tertiary' recycling.

The prospects for recycling techniques, and especially for the lower energy option of mechanical recycling, should improve as life-cycle considerations are taken into account when designing plastics products, and identification procedures are improved.

Energy recovery

This might be regarded as the most basic form of feedstock recycling. It recovers only the chemical energy within the plastic by incineration and use of the heat energy emitted. Energy recovery might be appropriate for plastics 'waste' that is unsuitable for one of the above techniques, but the material itself becomes a gaseous emission, and in some cases toxic or polluting components in the emissions (such as hydrogen chloride from PVC) may constitute an additional hazard. Energy recovery is the only current option for materials such as glass-reinforced polyester (GRP) since the glass cannot be separated from the resin. This is sometimes called 'quaternary' recycling.

SELF-ASSESSMENT QUESTIONS

1. Suggest reasons why plastics have generally low density, referring to Table 2.6 in your answer and suggesting which polymers should have the lowest densities. Check these suggestions against Table 9.2.

2. Explain why large molecules are most likely to result in a solid state and indicate how the size of simple organic molecules can be increased to produce solids.

3. Explain the meaning of the term 'glass transition temperature' and its significance in relation to the performance of
   (a) crystalline polymers;
   (b) amorphous polymers.
   Refer to common polymers in your answer.

4. Outline the main attractions and the main disadvantages of plastics in relation to the other major groups of materials, metals and ceramics.

5. Suggest why PVC (polyvinyl chloride) is the most widely used plastic in the construction industry.

6. Discuss the use of unplasticised PVC in window frames, giving recommendations for successful use.

7. Explain why elastomers have such high flexibility, giving recommendations for successful use as joint sealants.

8. Explain why glass fibres are incorporated into glass-reinforced polyester (GRP) and give recommendations for successful use of this composite in cladding panels.

9. Outline the possible hazards of plastics in fire. Give three instances where serious problems could occur and ways of reducing the risk. Give three examples where plastics can be used with minimal risk.

10. Describe the opportunities and the difficulties of recycling plastics. Suggest ways in which recycling could be extended in the future to a wider range of products.

10 Paints

Chapter summary

- Painting systems and constituents.
- Common types of paint.
- Painting specific materials.

10.1 Introduction

Paints are surface coatings, generally suitable for site use and marketed in liquid form. They may be used for one or more of the following purposes:

- To protect the underlying surface by exclusion of the atmosphere, moisture, fungi and insects.
- To provide a decorative easily maintained surface.
- To provide light- and heat-reflecting properties.
- To give special effects; for example, inhibitive paints for protection of metals; electrically conductive paints as a source of heat; condensation-resisting paints.

Painting constitutes a small fraction of the initial cost of a building and a much higher proportion of the maintenance cost. It is, on this basis, advisable to pay careful attention to the subject at construction stage. Furthermore, there are a number of situations in which restoration is both difficult and expensive once the original surface has failed and weather has affected the substrate; for example, clear film forming coatings on timber, and painting of steel. In other situations, access becomes more difficult later – for example, fascia boards become obscured by gutters. Such situations merit special care.

The environmental considerations are given in Table 10.1.

Table 10.1 *Environmental considerations: paints*

Consideration	Assessment
Raw material availability	Most paints are based upon organic binders derived from oil; they are a finite resource
Extraction	No specific problems
Energy used in manufacture	Processing energy contents are high but these are not significant in view of the small quantities normally used
Health/safety hazards	There may be hazards during manufacture, transport and use, especially when organic solvents or binders are used
Recyclability	Not normally possible since paint cannot be separated from substrate

10.2 Painting systems

There are usually three stages in a painting system – primer, undercoat and finishing coat – but undercoats and priming coats do not in themselves provide an impermeable dirt-resistant coating.

- **Primer**
  The function of the primer is to grip the substrate, to provide protection against corrosion/dampness and to provide a good key for remaining coats.
- **Undercoat**
  The function of the undercoat is to provide good opacity (hiding power) together with a smooth surface which provides a good key for the finishing coat. Undercoats usually contain large quantities of pigment to provide hiding power.
- **Finishing coat**
  This must provide a durable layer of the required colour and texture. Traditionally, most finishing coats were gloss and these tend to have the best resistance to dirt since they provide very smooth surfaces. Silk or matt finishes can be obtained if preferred. Some paint types such as emulsions will not normally give the high gloss of traditional oil paints.

Constituents of a paint

The types and proportions of paint constituents tend to be evolved by the manufacturer from experience rather than being designed from 'first principles'. Any one product will be subject to at least small modifications from time to time.

The main components of a paint are the vehicle or binder, the pigment and the extender.

Vehicle or binder

This is the fluid material in the paint which must harden after application. The hardening process may be due to one of the following:

(a) *Polymerisation by chemical reaction with air in the atmosphere.* Such paints tend to form a film in a part empty can. They include ordinary 'oil' paints.
(b) *Coalescence of an emulsion.* Emulsions are pre-polymerised into very small particles which are prevented from coalescing by an emulsifying agent. They set by water loss leading to 'breaking' of the emulsion.
(c) *Evaporation of a solvent.* Solvents need to be volatile, hence they are often flammable.

Paints based on types (a) and (b) are described as convertible coatings because once set, they cannot easily be re-softened. Once weathered, application of new coatings to these paints therefore relies on the previous paint being roughened to provide a key.

Paints based on type (c) are described as non-convertible since they can be re-softened by application of a suitable solvent. Subsequent coats also tend to fuse into previous coats and they do not form films in the can (though the paint may thicken by solvent loss due to evaporation).

The vehicle is largely responsible for the gloss and mechanical properties of the final coating. Vehicles may be blended with:

- **Driers**, which modify hardening properties.
- **Plasticisers**, which increase the flexibility of the hardened film.
- **Solvents**, which adjust the viscosity of the wet paint.
- **Other additives** – for example, with fungicidal action.

Pigments

These are fine insoluble particles which give the colouring ability and body to the paint. Primers and undercoats tend to have large proportions of pigment to produce opacity, while finish coats have low proportions, since, to produce a gloss, the pigment should be beneath the surface. The particle size of pigments is very small to enable maximum colouring power to be obtained by minimum thickness of material. Inorganic pigments, such as titanium dioxide (white), have the best performance in respect of resistance to solvents, colour fastness and heat resistance, though organic pigments tend to produce the brightest, cleanest colours.

Extenders

These can be added to control the flow characteristics and gloss of the paint with the added advantage of reducing the cost. Because they are not involved in the colouring process they have a particle size larger than that of the pigment.

10.3 Some common types of paint

Oil (alkyd resin) paints and varnishes

These are well established and are still the most widely used paints for general purposes, including painting of wood and metals. They were traditionally based on linseed oil, but modern oil paints are manufactured from alkyd, polyurethane, or other synthetic resins, which allow greater control over flow characteristics, hardening time and hardness/flexibility of the dried film. These paints comprise long-chain molecules rather like thermoplastics. They contain unsaturated groups (see Chapter 9) so that, on exposure to oxygen, oxygen atoms link the chains together and convert them to hardened films with the properties of thermosetting plastics. They become resistant to solution in oils from which they were formed. Unfortunately the hardening process continues slowly with time and these paints tend to become brittle over a period of years, especially if exposed to substantial levels of sunlight. This leads to cracking, especially when they are applied to substrates with high movement characteristics such as timber.

Saponification of oil-based paints and varnishes

Oil- or alkyd-based paints are made by reacting organic acids with alcohols such as glycerol. If an alkali such as calcium hydroxide contacts an oil-based film, there is a tendency to revert to glycerol with the production of the corresponding salt, which in this case is a soapy material – hence the name, 'saponification'. This leads to the breakdown of films and formation of a scum. Hence, oil-containing paints (including polyurethanes) should not be used on alkaline substrates such as asbestos cement, concretes, plasters or renders formed from Portland cements, especially when new or if there is a risk of dampness. Alkali-resistant primers, such as PVA emulsion paints, should be applied.

Emulsion paints

These are now very widely used in interior decorating. Many types are based upon polyvinyl acetate (PVA) emulsion and are suitable for application to new cement or plaster. The molecules are very large but are dispersed in water by colloids to give particles of approximately 1 μm in size. Hence, these paints have the advantage of being water-miscible, although, on drying out, coalescence of polymer particles occurs, resulting in a coherent film with moderate resistance to water (Fig. 10.1). The film is, however, not continuous, so that the substrate can, if necessary, dry out through the film. Acrylic emulsions for painting timber have now been produced, including a form which results in a medium gloss finish. Emulsions must have a certain amount of thermal energy to coalesce, hence, there is a 'minimum film formation temperature' (MFFT) for each type. Typical MFFT values are, for PVA, 7 °C, and for acrylic copolymers, 9 °C. They should not be used below these temperatures.

Fig. 10.1 *Coalescence of an emulsion to form a coherent but non-continuous film.*

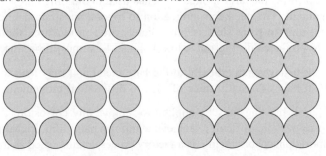

Emulsion paints offer the advantage that brushes can be washed out in water so that organic solvents, with their attendant hazards, can be dispensed with. However, brushes must be washed periodically in use since, once hardened, the paint film is resistant to dissolution in water and normal solvents.

Cellulose paints

These are solvent-based (non-convertible) paints. The cellulose constituent is in the form of nitro-cellulose dissolved in a solvent such as acetone. Plasticisers are added to give elasticity and synthetic resins to give a gloss, since pure cellulose gives little gloss. Drying usually occurs rapidly but well-ventilated areas are essential and the paint is highly flammable. Cellulose paints are most suited to spray application (though retarded varieties for brushing are available). An added attraction is that subsequent coats, even if applied long after earlier coats, bond into each other. Hence they adhere well.

Since they give off a penetrating odour, the use of cellulose paints tends to be restricted to factory application. In these conditions, high-quality finishes can be obtained and the resulting coat has good resistance to fungal attack and to chemicals, including alkalis.

The use of cellulose-based paints has now greatly reduced on account of the environmental hazards associated with the volatile organic solvents required.

Bituminous paints

These are intended primarily for the protection of metals used externally and have poor gloss-retention properties. They are amenable to application in thick coats which give good protection, though the solvents used sometimes cause lifting, if applied over oil-based paints, or bleeding, in subsequently applied oil-based coats. Sunlight softens the paint, though resistance can be improved by the use of aluminium in the final coat. Chlorinated-rubber paints have similar properties; some uses are based on their resistance to alkalis.

10.4 Painting specific materials

Ferrous metals

Steel forms the largest bulk of metals used in building and is one of the most difficult to maintain. The need for application to a good substrate cannot be overemphasised; the best time to paint steel is immediately after production, though mill scale (iron oxide film produced during hot rolling) should be removed because:

- It behaves cathodically to the bare metal and may lead to local corrosion.
- It may eventually flake off due to differential movement.

Grit blasting may be applied to remove any corrosion, though a very rough finish makes it more difficult to achieve a uniform paint film. **Pickling** – treatment with hydrochloric/phosphoric acid – is a factory process mainly used as a pre-galvanising treatment.

Inhibitive primers, which interfere with the corrosion process should water be present, include **red lead (lead oxide), zinc dust, zinc chromate** and **zinc phosphate**. Of these, red lead, though toxic, is still preferred in 'safe' situations because it is fairly tolerant of poorly prepared surfaces and is amenable to application in thick coats. **Metallic lead** primers, though non-inhibitive, are fairly tolerant of poor surfaces and may be flow more easily and dry more quickly than red lead. They also have superior chemical resistance. On account of their toxicity, lead-based primers are not recommended for use in domestic situations.

Where a decorative finish is required, **alkyd** or **aluminium paints** may be used. **Red oxide (micaceous iron oxide)** is a moderately effective primer and is used both in undercoats and finishing coats. Special ('prefabrication') primers about 15–20 μm thick are often applied to steel soon after production to afford weather protection prior to and during fabrication. These have good weathering resistance without the need for further paint coats. They are typically based on phenolic or epoxy resins and may contain etchants and inhibitors.

A relatively recent development is **powder epoxy** coatings. These are applied electrostatically (giving very uniform coverage), the coating being hardened by baking. They are factory applied to products which require good long-term protection. Reinforcing steels are available with such protection.

> The wetter the situation and the more aggressive the climate or atmosphere, the more coats should be given.

In extreme situations, or where extended life without maintenance is required, protection is only likely to be achieved by impregnated wrappings, bituminous or coal-tar coatings, thick, factory-formed films, or prior treatment, such as galvanising.

Non-ferrous metals

Zinc and aluminium are the non-ferrous metals most likely to require surface coatings and *each provides a poor key for paint*, unless surface treatment is first carried out. Zinc, in particular, reacts with most oil-based paints, forming soluble salts which reduce adhesion. Zinc should be degreased with white spirit, followed by roughening of unweathered surfaces with emery paper or etching treatment. Primers containing phosphoric acid are effective in this respect, and they often also contain an inhibitor, such as zinc chromate. Other suitable primers contain calcium plumbate, zinc dust or zinc oxide. For aluminium, etching is again an advantage, followed by application of zinc chromate or red-oxide primers. Lead-based primers are not suitable.

Wood

> Wood should be protected as soon as possible after the manufacturing process is completed since the surface is quite rapidly affected by weathering/ultraviolet light, as a result of which the paint adhesion properties significantly deteriorate.

Sanding, or planing where feasible, can restore the properties of a weathered surface to a suitable condition for painting. Preliminary treatment includes stopping holes and treatment of knots with shellac.

A primer is essential to penetrate and yet block the pore structure. Undercoats are unsatisfactory here, since they often do not penetrate the wood and may flake off later. Lead-based primers have been replaced by newer types such as aluminium (BS 4756) and acrylic water-borne primers (BS 5082), as well as the conventional 'solvent-based' primers. Acrylic primers are tolerant of higher moisture contents in the timber, though experience shows that permeability is too high for use in single coats as a protection for joinery timber exposed on site prior to installation. Where exposure of primed timber is a possibility, it is recommended that the specification should require that priming paints comply with the appropriate British Standard, since factory-based primers in particular, often do not comply with the 'six months' exposure test requirements of current standards. Undercoats contribute to the film thickness and, therefore, protection, though the final coat provides the bulk of protection. Alkyds and polyurethanes form the basis of most paints, though newer types, such as acrylic emulsions, are now available for external use.

Acrylic paints have demonstrated a number of advantages, including:

- Rapid drying
- Ease of application
- Ease of cleaning equipment and brushes
- Good durability, particularly cracking resistance.

They have the additional advantage that they obviate the need for organic solvents with their environmental implications.

The resistance to cracking mentioned above is due to the fact that setting does not involve oxygen; hence, embrittlement due to continued oxygen penetration does not occur. Possible problems of acrylic paints include the fact that they may inhibit hardening of fresh putty, they may be affected by rain during drying and gloss levels are not as good as those of conventional alkyd paints.

Microporous paints for wood

Some paints are claimed to be microporous or to accommodate wood movement or both. Tests on microporosity have shown that many such paints do 'breathe', though often not to a markedly greater extent than conventional paints. There is some evidence that, since cracking and blistering are less likely in such paints, they should have greater life. However, it should be appreciated that such paints would not be able to overcome high moisture contents caused by water admission at defective joints or breaks in the paint film. In some cases, there may be a possibility of higher resultant moisture contents in the wood, due to extra penetration through microporous paints during wet spells. Hence, although advantages may occur in some situations, the need for adequate preparation for painting, together with good maintenance, still applies. The microporosity of paints is also lost if more than three coats are applied and this should be considered when redecorating such coatings. Similarly, microporous properties will not be obtained if such paints are applied over conventional paints.

Such porous paints produce a 'sheen' rather than the high-gloss finish which may be preferred by some clients.

Varnishes and wood stains

Varnishes are essentially drying oils/resins with little or no pigment, which enhance the natural colouring and grain of timber. Formulations vary greatly, some being suited to external use, though the main problem with all varnishes is that exterior surface maintenance must be meticulous. Cracking or peeling of varnish very quickly leads to bleaching or staining of underlying surfaces due to exposure to water and/or ultraviolet light. Once affected, it is difficult to restore wood to its former state. The problem is especially severe on horizontal surfaces and south-facing aspects and it is not, therefore, recommended that varnishes be used in such situations.

When varnishes have microporous or other properties such that they are guaranteed not to flake or peel, they can be regarded as stains and may be treated as such. The essential features of stains are:

- They penetrate the wood so that texture and grain remain, though some ('high build') types form a surface film in addition.
- They have a low solids content.
- The coatings breathe quite freely, though water repellents are added to reduce admission of liquid water. Fungicides are also usually added to control mould growth.

The moisture contents of stained timber may rise from time to time, such that non-corrosive metal fixings should be used. Since many forms preserve the texture and sometimes the colour of underlying wood, imperfections may be seen and putty is not recommended for glazing – gaskets of glazing beads being preferred. Stains cannot cover cracks or gaps which might be covered by a paint, and any metal fixings will remain visible.

An important advantage of wood stains is that coatings generally erode rather than flake or crack, so that maintenance is confined to periodic washing and application of extra coats, though colours become progressively darker with age.

Low-build, transparent forms help to retain the natural character of the wood and may leave very little surface film so that protection to wood is correspondingly less and wood may begin to weather comparatively quickly. Many types are now available with coloured pigments which are becoming aesthetically more popular. These increase protection to the wood, though there is the risk that, if a thick surface film is applied, the failure characteristics of traditional timber varnishes, such as cracking and peeling, may occur. Higher-build varieties are recommended on 'difficult' woods, such as some hardwoods, with careful preparation, such as the removal of oils, if present. Adhesion is improved on rough-sawn wood finishes, especially of difficult timbers, though these have limited acceptability aesthetically at present.

Not surprisingly, with the variety of finishes obtainable and the relative ease of maintenance, the use of wood stains is rapidly increasing.

Plastics

Most plastics in common use do not require painting, and paint coats, once applied, cannot be removed by normal techniques. Paints, on the other hand, will reduce the rate of degradation of plastics such as polyethylene. Adhesion is poor unless the surface is first roughened to give a mechanical key. The impact strength of some plastics, such as PVC, may be adversely affected, if painted, by migration of solvent into the paint.

SELF-ASSESSMENT QUESTIONS

1. Describe the functions of the primer, undercoat and finishing coat in painting systems. Suggest what might be the effect on a wood substrate if
   (a) the primer is omitted;
   (b) the undercoat is omitted.

2. Explain the meaning of the term 'saponification', indicating the type of coating and situation in which it is likely to occur. State also how it can be avoided.

3. Write a specification for a protection system for structural steel in an exposed situation to which people have access.

4. Compare and contrast the following protection systems for wood:
   (a) conventional gloss paints;
   (b) microporous paints;
   (c) wood stains.

STANDARDS

BS 4756: 1971 (1991): *Ready mixed aluminium priming paints for woodwork*
BS 5082: 1993: *Water-borne priming paints for woodwork*

11 Bituminous materials

Chapter summary

- Types of bitumen and road tar.
- Principles of design of pavements.
- Asphalts and macadams.
- Surface dressings
- Defects in bituminous pavements or surfacings.
- Bituminous products for roofing.

11.1 Introduction

General usage of the term 'bituminous' covers products based on tar as well as those based on bitumen, though bitumen-based products are more widely employed in all forms of construction.

The bituminous group of materials offers the following general attractions:

- Excellent resistance to absorption/passage of water.
- Good adhesion to many materials.
- Resistance to dilute acids and alkalis.
- Good flexibility at normal temperatures.
- Properties can be varied to suit application.

Problems/disadvantages associated with bituminous materials are:

- They tend to become brittle at low temperatures and soft at high temperatures.
- They are subject to creep.
- They have low inherent stiffness and so must be carefully blended with harder materials where stresses are encountered.

The enviornmental considerations for bituminous materials are given in Table 11.1.

Table 11.1 Environmental considerations: bituminous materials

Consideration	Assessment
Raw material availability	Manufactured from oil, or lake asphalt – limited supplies
Extraction	No specific problems
Energy used in manufacture	Moderately high chemical energy required
Health/safety hazards	Oil transportation carries risks; vapours can be harmful
Recyclability	Possible in large-scale applications such as roads

11.2 Bitumen (BS 3690)

Bitumen consists essentially of hydrocarbons of varying molecular size which are soluble in carbon disulphide.

The larger molecular weight fractions are of solid nature and these are dispersed in the lower molecular weight fractions which are either of a resinous or an oily nature.

Bitumens occur naturally in the form of asphalts, which are mixtures of bitumen, minerals and water found in the form of rock or lake asphalt. The majority of bitumen is, however, produced as a residue from the fractional distillation of crude oil.

The properties of bitumen vary greatly according to composition which, in turn, depends on the manufacturing method and crude material. However, all bitumens are thermoplastic – they soften on heating, though they have no well-defined melting point. On cooling they tend to become progressively more brittle. They dissolve in many organic solvents, though this may not always be an advantage.

Production Process

The crude oil is initially heated in a distillation column to between 300 and 350 °C as a result of which volatile fractions vaporise and are removed, producing naphtha, kerosine and gas oil according to the position in the distillation column. The remaining material, known as the 'long residue', must be further processed to produce bitumen and this is done by heating in a partial vacuum in order to obtain further vaporisation without raising the temperature to over 400 °C, which would cause decomposition of the residue. The resulting 'short' residue may be modified by air blowing, carried out at a temperature of about 300 °C. The air has a number of effects, but overall the molecular weight of the 'asphaltenes' in the bitumen is increased and this reduces the susceptibility of the material to softening at high temperatures for a given 'penetration' (viscosity). The amount of air blowing can be varied, but road grades are usually subjected to a limited amount of this treatment.

Fig. 11.1 *Penetration apparatus.*

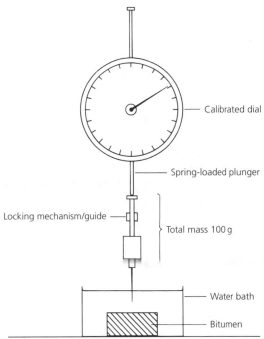

The essential feature of bitumens is that they are solid at ordinary temperatures and must therefore be heated prior to application, whether in roofing, flooring or roads. They are nevertheless widely employed since they have the advantage that hardening takes place immediately on cooling, so that there is only a minimal delay between construction and use. For grading purposes blown bitumens are described by their **penetration** and/or their **softening point**. Penetration is the distance that a standard shaped needle will penetrate a sample of bitumen when loaded in a standard manner (100 g load for 5 s at 25 °C; Fig. 11.1). The penetration is measured in units of 0.1 mm, e.g. '200 pen' bitumen means 20 mm penetration (see Experiment 11.1). Penetrations for most purposes lie in the range 30–300. The softening point temperature is found by a ring and ball test (Fig. 11.2; see Experiment 7.4). When bitumens have a penetration of less than, say, 100, the softening point test tends to give a more sensitive indication of hardness, the softening point increasing from about 45 °C for '100 pen' bitumens to over 60 °C for 'very low pen' bitumens. There is, however, no unique relationship between these two parameters and bitumen specifications often refer to both.

Cutback bitumens

These usually comprise straight-run bitumens to which a volatile fluxing oil such as kerosene or creosote is added in order to reduce viscosity. The handling and placing of products based on cutback bitumens can take place at much lower temperatures than straight-run bitumens. Subsequent hardening is by solvent evaporation, hence it is generally considerably slower than that of straight-run bitumens. The main application

Fig. 11.2 *Softening point apparatus.*

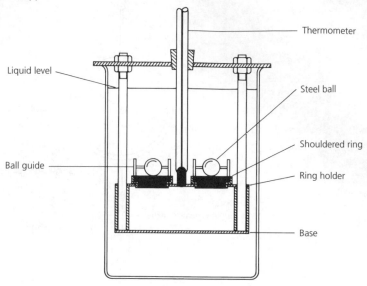

of cutback bitumens is in surface dressings. Cutbacks are graded using the standard tar viscometer in which the time in seconds for 50 ml of cutback to flow through a standard orifice at a standard temperature is measured (see Experiment 7.3 and Fig. 11.3). Values range typically from 50 to 200 s.

Bitumen emulsions

A bitumen emulsion contains minute bitumen particles – around 1 μm in size – dispersed in water by means of an emulsifying agent. These agents impart electric charges to the particle surface, thereby causing them to repel and prevent the formation of a continuous solid mass. Two types of emulsifier may be used: **anionic** emulsifiers, which impart a negative charge to the bitumen, and **cationic** emulsifiers which impart a positive charge.

The mechanism of solidification ('breaking') of emulsions commences when the emulsion contacts the mineral aggregate, which begins to inactivate the emulsifier by water absorption. In the case of cationic emulsions, breaking may be assisted by the neutralisation of the positive charge by the negative charges that are often present on solid mineral materials such as silica. Hence, cationic emulsions tend to break more quickly than ionic emulsions and may break without drying of the water. Anionic emulsions are nevertheless cheaper and more commonly used. Breaking is normally accompanied by a change in colour from brown to black.

Bitumen emulsions are widely used to produce damp-proof membranes, as curing membranes in concrete roadbases, as tack-coats in road surfacing and in surface dressings. They have the advantage of adhering to damp surfaces, though 'breaking' in such situations would normally be delayed. Thin films only must be used – ponding of an emulsion will result in very long drying times. Where thicker coatings, such as in damp-proofing, are

Fig. 11.3 *Section through standard tar viscometer.*

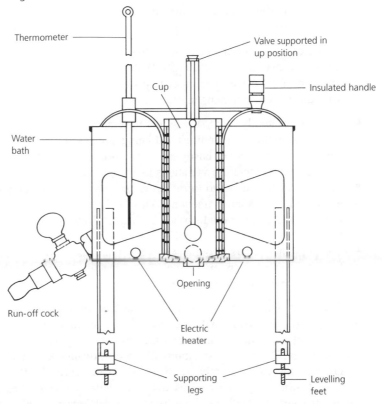

required, these should comprise several coats, each being allowed to dry before application of the next.

11.3 Road tar (BS 76)

Road tar is obtained by the destructive distillation of coal (or, less commonly, wood or shale) in the absence of air at temperatures up to 1000 °C.

The tar is driven off as a thick brown vapour, which is condensed and collected. The other products of distillation are gas, and oil fractions such as benzole.

The condensate is referred to as crude tar and this must be treated to produce road tars. Treatment comprises distillation at temperatures up to 350 °C in which light oil, naphthalene and creosote are progressively driven off. The remaining material (residue) is described as base tar or pitch and this is fluxed back with selected tar oils to give a road tar with the desired viscosity and distillation characteristics.

According to BS 76, there are two types of road tar: Type S and Type C.

- **Type S tars** contain more volatile oils and therefore tend to set more rapidly on the road, being used mainly in Surface dressings.
- **Type C tars** are less volatile and are intended for use with Coated macadam.

Tars are classified principally according to their viscosity as measured by the standard tar viscometer (Experiment 11.2). Water content is also important, BS 76 requiring a maximum of 0.5 per cent in order to prevent frothing on heating.

Tars differ from bitumens in the following main respects:

- They are generally more susceptible to temperature change, tending to soften quite quickly at higher temperatures and being subject to embrittlement at very low temperatures. Hence the viscosity of a tar chosen for a particular function will depend on seasonal temperature variations as well as traffic load requirements.
- Tars and pitches undergo 'weathering' in service – a combination of oxidation and evaporation of volatile fractions. This is an advantage in roads since smooth tar films, brought to the road surface by heavy traffic, weather away slowly and leave the aggregate exposed.
- The solubility of tars in organic solvents (such as petrol) is much lower than that of bitumen; hence tars are more suitable for surfacings which are subject to contamination by diesel or lubricating oils – such as parking areas for buses or commercial vehicles.

The main application of tar is as a surface dressing for roads, since it has the advantage over bitumens (with the exception of cationic emulsions) of superior adhesion in damp conditions. Dense tar surfacings, which are rather similar to dense bitumen macadams, are also used in situations where the advantages of a tar binder are required.

Overall, the use of tar has reduced since the early 1970s when North Sea gas replaced 'town' gas – the coal used for the latter being the main source of tar. An important source of tar is now the iron and steel industry which employs coke in the furnaces. The material is also obtained as a by-product of smokeless fuels.

Low-temperature tar

This term is misleading, since it may erroneously be taken to imply a tar for cold application. In fact the term refers to the method of production; a carbonising temperature of 600–750 °C being employed instead of 1000 °C, as for conventional tars. The lower temperature is used when coal is converted to smokeless fuels as distinct from the higher temperature 'destructive' distillation which was formerly used for the production of 'town' gas. This can be blended to give very similar properties to those of conventional tars. BS 76 makes provision for the use of such tars.

11.4 Principles of design of road pavements

Road pavements are designed to fulfil the following functions:

1. Provide a smooth skid-resistant surface.
2. Reduce the stresses produced by the wheels of vehicles to levels which can be withstood by the natural underlying material (subgrade).

Fig. 11.4 *Components of a road pavement together with principles of stress reduction.*

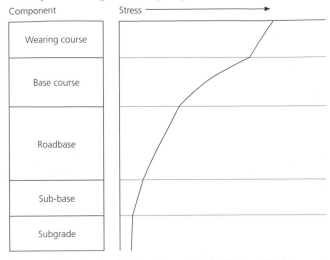

3. Protect the subgrade from water penetration and from frost (which would weaken the material).

Figure 11.4 shows the basic components of a road pavement.

- The **subgrade** is the natural material occurring at that position, though it may be adjusted to give the 'formation' level upon which the road sits.
- The **sub-base** will normally be a granular material, assisting in functions 2 and 3 above, as well as providing a working platform for construction traffic.
- The **roadbase** performs the main structural function. It may employ a cement binder, a bituminous binder, or (less commonly) simply crushed stone.
- The **surfacing** may be of concrete, or, more commonly, a bituminous material. The latter will be in two layers: a base course to provide a suitable platform for the other layer, the wearing course, which provides the wearing surface and normally sheds water from the road.

The stress at the road surface is approximately equal to the air pressure in the tyre; hence commercial vehicles, which operate at much higher tyre pressures than lighter vehicles, result in much higher surface stresses. The road pavement as a whole reduces wheel load stresses to acceptable levels in the subgrade, the rate of stress reduction depending on the relative stiffnesses of the various layers. Figure 11.4 shows schematically how stresses might be reduced. The subject is, however, quite complex, since excessive stress in any one layer must be avoided, the layers must work together and, most important, fatigue failure, resulting from repeated application of wheel loads, must be avoided. In addition, the layers must be able to withstand movements resulting from temperature or other effects.

Fig. 11.5 *Overall aggregate gradings for typical rolled asphalt and dense bitumen macadam. The rolled asphalt contains some coarse material and a high proportion of fine material. The macadam is composed mainly of coarse material.*

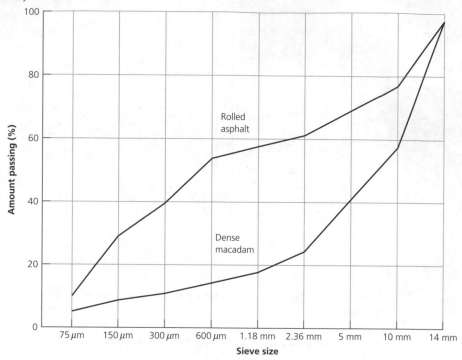

11.5 Asphalts and macadams

Asphalts are bituminous mixtures containing substantial amounts of fine material in the form of sand and filler. The sand and filler, together with the binder, form a stiff 'mortar' which provides strength and stiffness in the final product. They are described as 'gap graded' since there is little material between the sizes 2.36 and 10 mm. The particle grading is shown in Fig. 11.5, the curve being relatively flat between 2.36 and 10 mm. They employ a relatively hard bitumen in quite high proportions to produce an impermeable product with low void content and high durability

Macadams are materials which rely mainly on particle interlock rather than a stiff mortar to provide strength and stiffness. They normally contain a wide range of particle sizes in a continuous grading (Fig. 11.5), voids in the material being filled by particles of smaller size, as in concrete. Macadams coated with a bituminous binder are described as 'coated' macadams. The bitumen acts as a lubricant as well as a binder during compaction. Macadams use smaller quantities of bitumen than asphalts – typically 4–6 per cent, and hence are slightly cheaper than asphalts. Some macadams for roadbase construction may not have a bituminous binder, and are referred to as 'wet-mix' and 'dry-bound' macadams.

The principal forms of asphalts and macadams are as follows:

Fig. 11.6 *Hot-rolled asphalt containing 20 mm precoated chippings, rolled into a dense impermeable mortar. (The coin – a 10 p piece – gives the scale.)*

Hot-rolled asphalts (BS 594)

These comprise a mixture of aggregate, filler and 'asphaltic cement', which would generally be a 50 'pen' bitumen. They may be used in roadbases, base courses and wearing courses.

- **Roadbases** and **base courses** employ high stone contents – typically 50–60 per cent – and the aggregate is of relatively large size, such as 14 or 28 mm. They are laid in thicknesses up to 150 mm.
- **Wearing courses** contain higher binder contents and lower stone contents – typically 30 per cent – of smaller maximum size, such as 10 or 14 mm, and are laid 40–50 mm thick. Precoated chippings of size 14 or 20 mm are normally rolled in after laying to provide surface texture and, hence, skid resistance (Fig. 11.6).

The optimum binder content in a rolled asphalt is that at which maximum density and loadbearing capacity are obtained. They may be determined by experiment or, alternatively, 'recipe' type mixes can be used.

Hot-rolled asphalts have low permeability and high durability and are widely used on heavily trafficked roads and city streets.

Mastic asphalts (BS 1446; BS 1447)

These are high-quality surfacing materials with a relatively large binder content – between 6 and 20 per cent of a bitumen of low penetration, typically between 15 and 25. The aggregate may be natural rock asphalt (BS 1446) or, more commonly, limestone (BS 1447), each ground to a powder which largely passes a 2.36 mm sieve. Relatively high application temperatures in the region of 200 °C are necessary and the material is spread by hand – a process involving a good deal of skill, the material being applied in layers to a final thickness in the range 25–50 mm. The product contains less than 1 per cent voids and is impermeable. An important use is in road carriageways subject to the highest traffic intensities, applications including bridge decks, tunnels and deceleration areas. In these situations precoated chippings are rolled in to improve skid resistance.

A further important application is in flat roofing (see Section 11.8).

Bitumen macadams (BS 4987)

These are based on bitumens of 100–300 penetration, together with graded aggregates of size up to 40 mm depending on thickness. They have lower fines contents leading to a higher void content than asphalts (up to 25 per cent, depending on type). This reduces progressively due to compaction by traffic. Macadams have good workability due to their low fines content and are easier to lay than hot-rolled asphalts.

The texture of a macadam depends on the maximum aggregate size and on the actual content of fines – the amount of material passing a 3.35 mm sieve. There are several types:

Open-graded macadams

These have a maximum size of 10–14 mm and fines content not exceeding 25 per cent. They can be used as base courses (20 mm aggregate) and wearing courses (10 or 14 mm aggregate), though since they are permeable, an impermeable surface dressing would be needed when used as a wearing course.

Alternatively, the impervious layer can be positioned **underneath** the wearing course, which, when used in conjunction with an open-graded macadam of aggregate size 20 mm, leads to *pervious macadam*, greatly reducing spray in wet weather as well as giving a quieter ride (Fig. 11.7). In view of their safety advantages trials are currently being conducted on pervious macadams to determine suitability for heavily trafficked roads, though open-graded macadams are not generally considered suitable for the heaviest traffic densities.

Fig. 11.7 *Pervious macadam, showing the open texture, leading to free draining through the wearing course. The material shown here (15 mm) is widely used on French autoroutes.*

Medium-graded macadams

These have a maximum size of 6 mm with 45–65 per cent fines and can be used for non-trafficked areas such as footways.

Dense macadams

These have a fines content of about 37 per cent and typically 28 mm or 40 mm maximum size. The high fines results in low void contents of between 5 and 10 per cent, and are thus suitable for roadbases and base courses for heavily trafficked roads. The 1988 edition of BS 4987 refers to dense macadam wearing courses of aggregate size 10 and 14 mm, described as **close-graded** macadam, and introduces a new **dense wearing course** of aggregate size 6 mm, of lower void content. These materials are nevertheless not completely impervious and do not have the durability or skid resistance required for roads subject to the heaviest traffic loads.

Fine-graded macadam

This type of macadam (formerly called 'cold asphalt') is based on high penetration or heavily cutback bitumens and is designed to be used cold or warm. Fine and course varieties based on 6 and 10 mm aggregates respectively are available. Fine varieties are laid in thicknesses of about 20 mm and course varieties to 30 mm. They are used for regulating and patching purposes and for footpaths, having the advantage that they can be stored for some time (Fig. 11.8). They are initially soft and permeable but gradually improve in both respects on trafficking. They make little contribution to the strength of the pavement and will be indented by sustained point loads.

Fig. 11.8   *Fine-grained macadam (laid cold), used for lightly trafficked areas such as footpaths.*

11.6 Surface dressings

These are a quick, convenient and economic means of arresting surface deterioration of a road surface. The procedure involves cleaning off loose debris by brushing, spraying a bituminous emulsion or cutback bitumen onto the dry road surface, applying chippings to the wet binder, and rolling to embed the chippings. The choice of chipping size must take into account the existing surface texture and the hardness of the existing binder. An important part of the operation is to allow traffic to use the dressed surface, but with a strictly controlled maximum speed, to avoid wheel traction tearing the new surface. The traffic helps to embed the chippings. Surface dressing is only really feasible in summer months, when the temperature is high enough to promote drying/solvent loss, and it will not arrest a deteriorating road structure.

11.7 Defects in bituminous pavements or surfacings

Cracking

Cracking is likely to lead to serious problems in all types of surfacing. Cracked surfacings permit water penetration which may lead to weakening of foundations and hence reduced loadbearing capacity. The penetration of water will also accelerate deterioration due to the effect of frost and de-icing salts.

Cracking is caused essentially by movements which exceed those the materials can absorb. In flexible roads laid on cement-bound bases, a pattern of well-spaced cracks frequently forms in the bituminous surfacing, corresponding to cracks previously formed in the roadbase. These 'reflected' cracks can be minimised if, during construction, measures are taken which will result in a relatively large number of fine cracks in the base, so that movements at any one crack are reduced. Thicker or more flexible surfacings also reduce the extent of the problem. Once such cracking patterns are produced they are difficult to cure permanently but are relatively easy to seal and resurfacing will effect a short-term remedy.

Finer cracks or crazing may result when a bituminous surface is unable to absorb the thermal movements to which it is subjected. This may, in the first instance, be due to the use of a bitumen or tar having too low a penetration value, though a further cause is often overheating prior to use, which results in the loss of fluxing oils. Cracking may also result from overloading of the material either due to excessively high stresses caused by vehicle overloading or because the road foundation is inadequate. The latter may be the result of water penetration, for example, which increases stresses in the bituminous layers because of inadequate stiffness.

Deformation

Surface deformation in roads is the result of prolonged or severe mechanical stress. It is quite common in deceleration areas of roads such as roundabouts, traffic lights or bus stops, especially on downward gradients, being worst in the wheel tracks of heavily trafficked roads, especially in hot weather. The cause is the use of either too much binder, too soft a binder or incorrect aggregate grading and it leads to corrugations of the surfacing. These then exacerbate the problem as the upward-sloping sections of each undulation become subject to higher stresses under a given vehicle load. To affect a cure it is necessary to replace the surfacing with one designed to have the extra stiffness required.

A more serious form of deformation may occur in which the roadbase is also affected. This is caused by structural failure of the pavement as a whole and results in large depressions in the road surface, often in the wheel tracks of heavy vehicles in the 'slow' lane of multi-lane carriageways. In such situations the surfacing and roadbase must be removed and replaced to an uprated specification.

Fig. 11.9 *Varying degrees of embedment of chippings applied as surface dressing: (a) satisfactory – approximately half each stone covered; (b) too low, leading to 'scabbing'; (c) too high, leading to fatting-up.*

(a)   (b)   (c)

Embedment

Embedment is a term commonly used in the context of road surfacings and it describes the sinking of chippings into the binder and, in some cases, the underlying material. Particularly important is surface dressing where the binder and chipping size employed should be such that, after compaction by traffic, the lower half of each surface chipping is embedded (Fig. 11.9(a)). Where embedment is less than this (Fig. 11.9(b)), there will be a tendency for chippings to break loose ('scabbing'). Conversely, excessive use of binder, combined with soft substrate, may lead to total embedment and hence to 'fatting-up' with loss of surface texture and skid resistance (Fig. 11.9(c)). Where this occurs, a further surface dressing can be applied with relatively low binder content, together with increased chipping size (up to 20 mm) in order to result in the correct degree of embedment after compaction by traffic.

Embedment can also lead to loss of surface texture and hence high-speed skid-resistance in hot-rolled asphalts.

11.8 Bituminous products for roofing

Bitumens have found application for roofing applications for many years, and where quality materials are combined with good practice, satisfactory durability can be obtained at competitive cost.

In the UK, bitumen products form a part of most roofs since sarking felt is widely used to provide an additional barrier to wind-blown moisture in a pitched roof, and in flat roofs the bitumen is totally responsible for the waterproofing function.

Pitched roofing

The felt is held in place by tiling battens, the layer providing a quickly applied protective covering which assists in the construction process. Sarking felt has quite specific requirements; it does not need to be highly impermeable since any moisture deposited is quickly shed downwards. However, it must have adequate strength to withstand the stresses imposed during fixing, wind pressures and its own weight in the long term. This is achieved by a coarse mesh fabric reinforcement. Correct procedure in sarking felt for pitched roofs is to continue the felt over the fascia at the eaves into the gutter, and it must be properly supported at the position to avoid ponding of moisture and eventual failure. This would lead, in turn, to dampness and wet rot in the fascia and soffit boards.

Flat roofing

Bituminous felts have traditionally been associated with lower cost flat roofing applications. Flat roofs are normally cheaper to build than pitched roof equivalents and built-up felt systems contribute significantly to cost savings. The basic system involves building up several (usually three) layers of material to form a watertight unit, successive layers being bonded together with hot applied bitumen. The multilayer system reduces the risk of water penetration if defects occur, for example, in individual sheets. It also permits easy production of a membrane of suitable total thickness.

The durability of such systems has been found to be variable. In virtually all cases replacement is called for within building lifetimes, though their cost is so competitive that this scenario is generally acceptable. Very short lifetimes are likely to result from use of inappropriate materials or techniques. A summary of some of the considerations is given here to focus attention upon the critical aspects of performance.

Felt quality

All felts require some form of reinforcement to withstand stresses arising from execution of the work and movement in service. Some cheaper early felts were bonded by organic (vegetable) fibres (class 1) which deteriorated relatively rapidly on exposure. These, together with asbestos-based felts (class 2) no longer form a part of BS 747. Two classes are currently specified for built-up felt roofing in BS 747:

- Glass fibre base   Class 3
- Polyester base   Class 5

Of these, polyester felts have a better track record, especially in warmer climates. Polymer modified versions are available offering the option of torch application and better fatigue resistance at higher cost.

Design for movement

Cracking in built-up felt roofing is generally the result of fatigue, in which repeated movements eventually overstress the material. It will be recalled (Chapter 1) that fatigue resistance is measured in cycles. Each diurnal temperature change can be regarded as one cycle. A 10-year life would therefore require resistance to about 10×365 or 3650 cycles. Clearly, in practice, the damaging effect of such cycles would depend on how well the roof is protected from direct sunlight and on how deep the temperature cycles are.

It has been found that well-insulated roofs can cause special problems because the depth of cycles may be increased. For example, in winter weather, which is the time at which the material has least flexibility, surface temperatures of a well-insulated roof can be significantly lower than a poorly insulated roof which is kept warm by heat losses.

Felt may be rigidly bonded to a concrete substrate, but when applied to timber some movement provision should be incorporated and this is achieved by nailing only (not bonding) the first layer. Adhesive bonding to foam insulating boards can lead to rapid fatigue failure at joints. Where movement occurs, stresses will be highest where the material is of thinnest section. Figure 11.10 shows a mineral-surfaced felt failure at the eaves. The covering had achieved over 10 years service, but the mineral surface had

Fig. 11.10 *Fatigue cracking of weathered mineral felt near to a verge.*

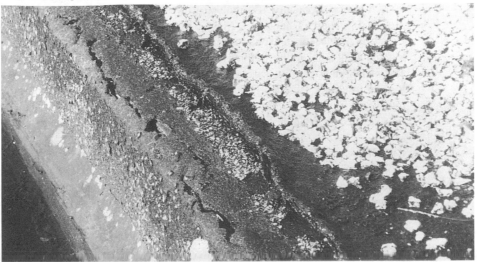

weathered away and caused fatigue failure in the embrittled material adjacent to a reduction in thickness.

Workmanship

This forms a crucial part of a successful system. Possible problems include:

- Incorrect detailing – for example, failing to treat edges and upstands correctly.
- Operating in wet weather, leading to moisture being trapped within the layers.
- Overheating the bitumen, leading to loss of plasticisers.
- Poor bonding at joints, leading to water penetration at joints.
- Failure to protect the final product from solar radiation.

Fall of roof

Since no felt is completely waterproof, the roof must be laid to a sufficient fall to avoid all ponding. The fall required therefore depends on construction tolerances, but given the risk that joists in timber roofs may contain distortion defects, design falls should veer towards higher rather than lower values. Minimum falls of 1 in 80 are quoted in codes of practice (e.g. BS 6229) though to achieve these in the long term the mean fall would probably need to be twice this value. Hence a fall of 75 mm in 3 m (1 in 40) should suffice for most purposes.

Asphalt roofing (see also section 11.5)

The asphalt is laid in two layers on isolating felt to allow movement and is held in place by its own weight. The substrate for the asphalt should be rigid – for example, concrete, rather than timber, since asphalts tend to crack under excessive flexing, especially after ageing. Solar reflective treatments help to reduce embrittlement caused by ageing.

Mastic asphalt roofs provide much better resistance to pedestrian traffic than built-up felt roofs, especially if suitably gritted. They will also withstand vehicle traffic provided support is adequate. Mastic asphalt is also used for flooring and tanking. Cracking may occur in asphalt roofs where the isolation membrane between the surfacing and deck is absent or ineffective. Joints in decking, together with dressings, flashings and weatherings require careful design and execution to avoid excessive movements leading to cracks. Where a few well-defined cracks exist, it may be possible to cutback, make provision for movement and apply fresh asphalt. Where significant water penetration has occurred, however, complete replacement will be necessary.

Crazing

This is a surface effect and does not, initially, lead to water penetration though it accelerates overall failure of the material, since cracks concentrate the tensile stresses occurring when cold weather causes thermal contraction. Most crazing occurs in cold weather, since this results in tensile stress at a time when the tensile strain capacity of the material is at its minimum. Surface crazing may be at least temporarily alleviated by means of surface treatments, which also require protection from solar radiation. Deeper crazing, resulting in moisture penetration, will necessitate replacement of the material.

Blistering

This is caused by the presence of moisture under the surface. The blistering occurs in hot weather when heat vaporises the water, which swells the softened mastic. Frequently the blisters remain on subsequent cooling – for example, if rain stiffens the asphalt and prevents resetting. Blisters are initially waterproof but may eventually leak in cold weather if damaged mechanically. The asphalt is also invariably thinner at these positions, increasing vulnerability. Minor blisters may be left, or patched in dry weather. Severe blistering is indicative of a design or construction fault which calls for replacement. The problem occurs most commonly with 'wet' substrates such as concrete which are not effectively isolated by a membrane from the asphalt. By the use of such a membrane, together with dry conditions during construction, blistering should be avoidable.

EXPERIMENTS: TESTS ON BITUMEN/TAR BINDERS

Introductory comment

These tests are based on British Standard specifications, although some have been abbreviated or simplified to take account of the limited time usually available. Further simplifications can be undertaken where only the principle of the test is being illustrated, though the experimenter should refer to the text of BS specifications where an accurate and valid test result is required.

Experiment 11.1 Determination of the penetration of bitumen (BS 2000: Part 49)

This test is used to determine the consistency of penetration grade 'straight-run' and stiffer 'cutback' bitumens at the standard temperature of 25 °C.

Apparatus

Penetration apparatus, as shown in Fig. 11.1. The dial gauge should be capable of reading to 0.1 mm and the mass of the needle and spindle assembly should total 100 g. Needles should be of hardened stainless steel, of 1 mm diameter with a 9° taper at one end to a truncated tip 0.15 mm in diameter. Sample containers should be of diameter 55 mm and internal depth 35 mm for penetrations less than 200 or diameter 70 mm and internal depth 45 mm for penetrations between 200 and 350. They may be of metal or glass. A water bath capable of maintaining a temperature of 25 °C ± 0.1 °C, stop-watch and thermometer are also required.

Preparation of specimens

Heat the sample until just sufficiently fluid to pour, stirring to avoid local overheating. Stirring also helps to remove air bubbles which may be present in blown bitumens. Avoid prolonged heating (over 30 min). Pour into the sample container judged to be appropriate to the original consistency of the material, tapping to remove air. The container should be almost full. Allow to cool for 1–2 h, according to container size and then insert into the water bath containing at least 10 litres of water and leave for a similar time to allow the temperature to reach 25 °C.

Procedure

The test should be carried out on the specimen while submerged in the water bath or while submerged in a suitable smaller vessel containing water at 25 °C. In each case check that the sample is firmly mounted and unable to rock. Lower the clean penetrometer needle until it just touches the bitumen. This may be facilitated by using the shadow of the needle or its reflected image. Take the reading of the dial gauge. Release the needle and start the stop-watch. After 5 s note the reading of the needle. The penetration is equal to the difference between the two readings expressed in units of 0.1 mm (for example, 20 mm penetration means '200' penetration bitumen). Make three determinations, the simplest method being to use a fresh clean needle at a different part of the surface for each determination. Find the average reading.

Experiment 11.2 Determination of the equiviscous temperature (e.v.t.) of tar using the standard tar viscometer (BS 76)

This test is designed for softer bituminous materials such as cutback bitumen. The e.v.t. is the temperature at which the viscosity of the material is of a standard value.

Apparatus

Standard tar viscometer (Fig. 11.3). The essential features are as follows:

- Tar cup, in the form of an open-topped brass cylinder with an accurate opening of diameter 10 mm at the base. This opening can be closed by a sphere attached to an operating rod.
- Water bath, which contains an accurate central sleeve into which the tar cup fits. It has three legs which are adjustable for levelling, a run-off cock, electric heater and stirring vanes. The latter are attached via a further concentric cylinder to a

curved upper shield provided with a handle for rotating. This shield also provides support for the tar cup valve.

- A 100 ml graduated measuring cylinder, thermometer and stop-watch are also required.

Sample preparation

Immerse a container of the sample in water at a temperature of not more than 30 °C above the expected equiviscous temperature. If no information is available as to the expected e.v.t., a preparation temperature of 65 °C may be used as a starting point, heating further as necessary to obtain adequate fluidity. BS 76 requires the warm sample to be strained through a 600 μm sieve to remove any material that may interfere with flow through the standard orifice. Stir the tar and keep covered while hot to avoid loss of volatile oils.

Procedure

Adjust the water bath so that the upper rim of the sleeve is horizontal, using the spirit level and the adjusting screws on the legs. Fit the stirrer assembly and fill the water bath to a level 10 mm from the top. Place the tar cap and valve in the sleeve.

The e.v.t. should be 'guessed'; if no information is available, a temperature of 35 °C may be taken. The temperature in the water bath should then be adjusted to 0.05 °C higher than the required temperature and maintained within 0.05 °C of this value (that is, between 0 and 0.1 °C higher than the required temperature) throughout the test. Stir the water frequently.

Pour 250 ml of the tar at the temperature of preparation into a beaker and allow to cool to 0.5 °C above the required test temperature. Close the orifice in the tar cup and pour tar into the cup until the correct level (indicated by a peg) is reached. Place the graduated measuring receiving cylinder under the tar cup and add 20 ml of soap solution to this cylinder to help prevent tar sticking to it. Stir the tar gently, using a thermometer, until the temperature is at the required test value and then withdraw and remove any excess tar from the thermometer. Lift the valve clear of the tar and commence timing when the meniscus of the soap reaches the 25 ml mark. Stop timing when the 75 ml mark is reached. Replace the valve. Empty the receiver to avoid the tar sticking to the sides.

The time recorded will be that for 50 ml of tar (discounting the first 25 ml) to flow from the cup. At the equiviscous temperature, the time taken would be 50 s. For times in the range of 33–75 s, Table 11.2 will enable the correct e.v.t. to be obtained. Suppose, for example, a time of 65 s was obtained at a test temperature of 35 °C. The correction from the table is +1.6 °C, and so the e.v.t. value is 36.6 °C. If the time obtained is outside the values indicated in heavy type in the table, the experiment should be repeated using Table 11.2 as a guide to select a new test temperature.

Typical values of equiviscous temperature for tars

Surface dressing	Heavy traffic	42–46 °C
	Average roads	38–42 °C
	Light traffic	34 °C

Viscosity (s)	0	1	2	3	4	5	6	7	8	9
Table 11.2 *Corrections to equiviscous temperatures for tar*										
10	−10.4	−9.8	−9.2	−8.7	−8.2	−7.7	−7.3	−6.9	−6.5	−6.1
20	−5.7	−5.4	−5.1	−4.8	−4.4	−4.3	−4.0	−3.8	−3.5	−3.3
30	−3.1	−2.9	−2.7	**−2.5**	**−2.3**	**−2.2**	**−2.0**	**−1.9**	**−1.7**	**−1.5**
40	**−1.4**	**−1.2**	**−1.0**	**−0.0**	**−0.8**	**−0.6**	**−0.5**	**−0.4**	**−0.3**	**−0.1**
50	**0**	**+0.1**	**+0.2**	**+0.3**	**+0.5**	**+0.6**	**+0.7**	**+0.8**	**+0.9**	**+1.0**
60	**+1.1**	**+1.2**	**+1.3**	**+1.4**	**+1.5**	**+1.6**	**+1.7**	**+1.7**	**+1.8**	**+1.9**
70	**+2.0**	**+2.1**	**+2.2**	**+2.3**	**+2.3**	**+2.4**	+2.5	+2.5	+2.6	+2.7
80	+2.8	+2.8	+2.9	+3.0	+3.0	+3.1	+3.1	+3.2	+3.3	+3.3
90	+3.4	+3.5	+3.4	+3.6	+3.6	+3.7	+3.7	+3.8	+3.8	+3.9
100	+4.0	+4.0	+4.1	+4.1	+4.2	+4.2	+4.3	+4.3	+4.4	+4.4
110	+4.5	+4.6	+4.6	+4.7	+4.7	+4.8	+4.8	+4.8	+4.9	+5.0
120	+5.0	+5.1	+5.1	+5.2	+5.2	+5.2	+5.3	+5.3	+5.4	+5.4
130	+5.5	+5.5	+5.5	+5.6	+5.6	+5.7	+5.7	+5.7	+5.8	+5.8
140	+5.9	+5.9	+6.0	+6.0	+6.0	+6.1	+6.1	+6.1	+6.2	+6.2

Note: That part of the table giving corrections for viscosities between 33 and 75 s inclusive (indicated by bold type) may alone be used in calculating the e.v.t. to be reported. The remainder of the table will be useful in ranging tests.

Coated materials	Dense roadbase (heavy traffic)	54 °C
	Base course, estate roads	38 °C

Experiment 11.3 Determination of the viscosity of cutback bitumen (standard tar viscometer BS 2000: Part 72)

The apparatus for this test is basically the same as that for Experiment 11.2, but there are some differences in procedure. The cup size should generally be 10 mm and the test temperature 40 °C, though a 4 mm cup and 25 °C temperature may be used for very fluid samples.

A trial test may be necessary to determine the correct cup size and temperature, although most cutbacks have viscosities in the range 50–200 s.

Procedure

According to BS 2000, the temperature of the cutback should be adjusted to within 0.1 °C of the test temperature by the use of a separate water bath. This requires the orifice of the viscometer to be closed by a cork inserted from beneath and the thermometer and valve to be located by a second cork with a central hole (for the thermometer) and slot fitted into the top. The cutback should be free from water and should first be heated to 60 °C for a period given below, depending on sample size, taking care to avoid loss of volatile constituents.

Vol of sample (ml)	Period of heating (hours)
750 ± 250	2 ± 0.25
1500 ± 500	2.5 ± 0.5
2500 ± 500	3 ± 0.5

Cool the cutback to the approximate test temperature. After fitting the corks, fill the viscometer cup with the cutback until it is level with the peg. Immerse in the separate water bath for 1.5 h at the correct temperature.

Set up the viscometer water bath and adjust to the correct temperature, stirring frequently. Transfer the viscometer cup to the viscometer water bath and remove the thermometer and corks. Check that the level of the cutback is correct. Time 50 ml of cutback through the orifice, as described in Experiment 11.2, except that light mineral oil is used in the graduated receiver instead of a soap solution. Hence, obtain the viscosity of the cutback expressed in seconds.

Experiment 11.4 Determination of the softening point of bitumen using the ring and ball method (BS 2000: Part 58)

This test is designed to be used with 'straight-run' (penetration grade) bitumens.

Apparatus

The apparatus enables tests to be carried out on two samples of the material. It comprises a support frame 25 mm above a lower plate, together with two tapered rings, two ball-centring guides and two steel balls (Fig. 11.2).

A heat-resistant glass container, 300 μm sieve, electric heater, mechanical (electric or magnetic) stirrer, thermometer and knife are also required. Distilled water is normally suitable as the heating medium; glycerol/dextrin debonding mixture (or a suitable grease).

Sample preparation

Heat the sample until fluid; with the exception of low penetration bitumens, a temperature of 130 °C should suffice. Stir until free of air bubbles and filter, if necessary, through a 300 μm sieve (though this should normally not be necessary). Heat the tapered rings to about the same temperature and place on a metal plate coated with a debonding mixture of glycerol and dextrin. Fill the rings with a slight excess of the liquid bitumen but avoid overflowing. Cool in air for 30 min and cut away the excess with a warm knife.

Procedure (for softening points of 80 °C and below)

Assemble the apparatus as in Fig. 11.2, with ball guides in position, and fill the glass container to 50 mm above the upper surface of the rings with freshly boiled distilled water at a temperature of 5 °C. After 15 min place the steel balls, which should be at a similar temperature, in the ball guides. Heat the bath, stirring constantly so that the temperature rises at 5 ± 0.5 °C per minute. Record the temperature at which each sample surrounding the ball touches the bottom plate. The temperatures obtained from the two samples should be within 1 °C of each other. Obtain the average temperature – this is the softening temperature.

SELF-ASSESSMENT QUESTIONS

1. Distinguish between bitumen and tar
   (a) in terms of manufacture;
   (b) in terms of properties/uses.

2. Comparing bitumen with tar, comment on
   (a) the weathering properties;
   (b) the temperature susceptibility;
   (c) the action of organic solvents.

3. Distinguish between blown bitumen and cutback bitumen. Give basic properties and applications of each.

4. Explain what is meant by a bitumen emulsion and describe briefly how it 'breaks' or sets. Hence indicate why ponding should be avoided. Give two applications of bitumen emulsions in construction.

5. Distinguish between asphalts and bitumen macadams as used in road construction. Give the special properties and applications of mastic asphalts.

6. Suggest possible causes of cracking in asphalt roofs and indicate how, by good design and construction, the risk of cracking can be minimised.

7. Distinguish between cold asphalt and dense bitumen macadam in respect of composition and properties.

8. Explain what is meant by blistering in asphalt or built-up felt roofs. Suggest
   (a) how the risk of blistering can be reduced;
   (b) what remedial steps may be taken.

9. Explain the meaning of the term 'embedment' in relation to surface dressing on roads. Indicate what circumstances would lead to 'fatting up' and 'scabbing' in surface dressings.

10. Describe tests which could be used for measuring
    (a) the hardness of a bitumen;
    (b) the behaviour of a bitumen at elevated temperatures.
    Explain the relationship between these properties.

REFERENCES

Road Note 39: *Recommendations for surface dressing*; Department of Transport; Transport and
   Road Research Laboratory, 1981
The Shell bitumen handbook 1990. Shell, UK Ltd

STANDARDS

Binder
BS 76: 1974 (1994): *Specification for tars for road purposes*
BS 3690: Part 1: 1989 (1997): *Specification for bitumens for roads and other paved areas*

Mixtures
BS 63: Part 1: 1987 (1994): *Single-sized aggregate for general purposes*
BS 594: *Hot rolled asphalt for roads and other paved areas*
   Part 1: 1992: *Specification for constituent materials and asphalt mixtures*
   Part 2: 1992: *Specification for transport, laying and compaction of rolled asphalt*
BS 1446: 1973 (1990): *Specification for mastic asphalt (natural rock asphalt fine aggregate)*
   for roads and footways
BS 1447: 1888: *Specification for mastic asphalt (limestone fine aggregate) for roads, footways*
   and pavings in building
BS 4987: *Coated macadams for roads and other paved areas*
   Part 1: 1993: *Specification for constituent materials and for mixtures*
   Part 2: 1988: *Specification for transport, laying and compaction*

Tests
BS 2000: *Methods of test for petroleum and its products*
   Part 47: 1993: *Solubility of bituminous binders*
   Part 49: 1993: *Determination of needle penetration of bituminous material*
   Part 58: 1993: *Determination of the softening point of bitumen. Ring and ball method*
   Part 72: 1993: *Determination of viscosity of cut-back bitumen*

Roofing felts
BS747: 1994: *Specification for roofing felts*

Answers to numerical self-assessment questions

Chapter 3

8. The sand complies with the medium and fine gradings of BS 882 (1992).

10. (a) 5.1 per cent
(b) 9.3 per cent.

11. C.F. incorrect.

13. 1(a) 0.47   1(b) 0.58   1(c) 0.52
2(a) 0.58   2(b) 0.47   2(c) 0.42
3(a) 0.53   3(b) 0.57   3(c) 0.46.

14. Water   9 litres
Cement   19.5 kg
Sand   29 kg
Coarse aggregate 61 kg.

15. Water   1.4 litres
Cement   3.5 kg
Sand   4.4 kg
Coarse aggregate 16.3 kg.

16. Mean $= 25.2$ N/mm^2
Standard deviation $= 3.6$ N/mm^2
Characteristic strength (2 per cent failures) $= 17.8$ N/mm^2
The required characteristic strength is not being reached.

17. The strength requirement of 25 N/mm^2 with 5 per cent permissible failures is being satisfied.

18. Water/cement ratio $= 0.52$.

Chapter 4

4. The brick is of engineering 'A' classification.

14. Porosity $= 67$ per cent.

15. Solid density $= 2364$ kg/m^3.

Chapter 7

6. Carbon equivalent = 0.473 per cent.

Chapter 8

8. (a) Reject   (b) GS   (c) Reject
   (d) SS   (e) GS   (f) GS
   (g) GS   (h) GS   (i) SS.

10. (a) Actual stress = 3.67 N/mm^2 (satisfactory)
    (b) Deflection = 4.6 mm (within permissible limits).

11. Applied stress is 0.44 N/mm^2. Maximum allowable stress is 1.6 N/mm^2. Hence the partition is of sufficient strength.

Additional reading

Building Research Establishment Digests

Addleson, L. and Rice, C. (1991) *Performance of Materials in Building*. Butterworth Heineman

Dinwoodie, J.M. (1989) *Wood, Natures Cellular Polymeric Fibre Composite*. Institute of Metals.

Doran, D.K. (1992) *Construction Materials Reference Book*. Butterworth Heineman

Gordon, J.E. (1976) *New Science of Strong Materials*. Pelican

Illston, J.M. (ed.) (1994) *Construction Materials, their Nature and Behaviour*. E & FN Spon

Jackson, N. and Dhir, R.K. (1988) *Civil Engineering Materials*, Macmillan

Murdock, L.J., Brooks, K.M. and Dewar, J.D. (eds) (1991) *Concrete Materials and Practice*, 6th edition. Arnold

Lyons, A.R. (1997) *Materials for Architects and Builders*. Arnold

Watson, K.L. (1975) *Materials in Chemical Perspective*. Stanley Thomas

Index

acidic corrosion of metals 205–8
acrylic
  paints 304
  plastics 287, 293–4
acrylonitrile butadiene styrene (ABS)
    287, 293–4
admixtures for concrete 90–93
aggregates for concrete
  batching 54
  bulking of 56
  density 55–6
  grading 52–5
  high density 56–7
  lightweight 56–7
  moisture in 57–60
  quality 57
  shape 65
  size 51–2
  surface texture 65
Agrement, British Board of 6
alkali silica reaction (ASR) 87–9
alkyd paints 300–301
aluminium 25, 202–3, 213
  painting of 303
  paints 303
amorphous materials 33
asbestos health risk 17
asphalts 314–16
atom
  electron shells 25–8
  structure of 22–32
  nucleus 22–5
atomic bonding
  covalent 29–31
  ionic 29
  metallic 31–2
  strengths 32–3, 34
  van der Waals 32, 280
austenite 192, 211
austenitic stainless steel 211–12

barium carbonate in bricks 118
bitumen 17, 307–11
bituminous materials
  emulsions 310–11
  mastics 315–16

paints 302
road pavements 312–19
roofing 319–22
tar 311–12
tests on 309–11
blast furnace cement 46
blocks 129–31
bonding of atoms 28–34, 276–80
boron diffusion to treat timber 233
brass 202–3
  dezincification 208
bricks
  calcium silicate 127–9
  clay 13, 117–27
  compressive strength 121–2
  concrete 129
  engineering 124–5
  Fletton 117
  frost resistance 125–7
  stock 118
  water absorption 122
British Board of Agrement 4
British Standards 2

calcium chloride in concrete 92–3
capillarity 142–4
carbon
  in iron 190–94
  in plastics 276–9
carbonation of concrete 89
cast iron 202–3, 212
cavity fills 146–7
cementite 190
cements
  composition 43–5
  controlled fineness 47
  effect of water/cement ratio 66
  fineness 44
  high-alumina 50–51
  hydration 50
  low-heat 46
  manufacture 43
  masonry 136–7
  MDF 35
  Portland 43–50
  Portland blast furnace 46

pozzolanic 46, 87
rapid hardening 45
replacement materials 47
setting time 47–9
standards 42
strength 49–50
sulphate-resisting 45, 87
tests 47–50
unsoundness 49
white 46
chipboard 257–9
chlorides in concrete 89
chlorofluorocarbons 17
clay bricks 117–27
combustion modified high resilience
    (CMHR) foams 294
concrete
  accelerators 92–3
  admixtures 90–93
  aerated 94–5
  air-entrained 93
  alkali silica reaction (ASR) 87–9
  autoclaved 95
  carbonation 88–9
  chlorides in 89
  cohesion 65
  compaction 61
  corrosion of steel in 88–90
  creep 199
  crystallisation damage 87
  curing 80
  durability 66–7, 85–8
  fresh density 79–80
  fresh, tests on 61–6
  frost resistance 95–6
  ground granulated blast furnace slag
    (GGBFS) in 46
  high-strength 95–6
  lightweight 93–6
  microsilica in 46
  mix design 60, 68–77
  no-fines 95
  plasticisers 66, 91–2
  prestressed 199–200
  recycling 98–100
  reinforced 196–9

retarders 92
segregation 65, 93
shrinkage 83–5
specification 67–8
statistical interpretation of cube results
  81–3
strength 66
sulphate attack 86–7
superplasticisers 92
surface finishes 96–8
trial mixes 77–9
water for 60
water reducers 66, 91–2
workability 61–5
conversion
  of high-alumina cement 51
  of timber 226–7
copper 202–3
corrosion
  in steel 88–90, 209
  in metals 205–12
covalent bond 29–31, 276
creep
  definition 12
  in concrete 199
cross-linking in polymers 280–81,
  284
crystal structure
  metals 33
  polymers 33
crystallisation damage in brickwork
  127
crystallites in polymers 282–3

damp, rising 148–50
density, bulk/solid 141–2
diamond 33
differential aeration, corrosion by
  209
durability, definition of 2, 16

effloresence in bricks 122–3
elastomers 288–9
electrode potential of metal 207
electrons 23–32
elements 25–7
embodied energy 18
emulsion paints 301–2, 304
energy considerations 18
environmental issues 18, 19–20
epoxide (epoxy) resins 290–91
Eurocodes 4
European standards 4

fatigue, definition of 13
fibre building boards 257–8
fibrous composites 36–40
fineness of cements 44–5
fire
  nature of 14
  retardant foams 242

formaldehyde gas 17
frost, effect on
  clay bricks 125–7
  concrete 85–6
fungal attack on timber 13

glass fibres 13, 37
glass transition temperature 283–4
glass
  manufacture 164–5
  laminated 167
  leaded lights 172
  strength of 165–6
  safety 166–8
  thermal performance 169–71
  toughness of 11
  wired 167
glulam 241, 259–60
granite aggregate for concrete 51,
  96
graphite 33
gunmetal 208
gypsum 37
gypsum plaster 174–83

hardboards 257–8
hardening of steel 191–3
hardness of metals, definition of 12
heat treatment of steel 191–3
high-alumina cement 50–51
hygroscopic salts 149

impact properties of steel 193–4
iron 25
ISO standards 6

joints in masonry 288–9

kitemark 8

lead 17, 202–3, 212
lead-based paints 302–3
lightweight aggregates 51, 56–7, 94
limes in mortar 136–7
linseed oil in paints 300–301
low-heat cement 46

macadams (bitumen) 314–17
manganese in steel 191, 195
martensite 192–3
masonry
  dampness in 141–50
  recycling of 151–2
  strength of 138–9
masonry cement 136–7
mastic asphalts 315–16
metals
  corrosion 16, 205–12
  crystals 186–90
  dislocations 188–9
  oxidation 205

recycling 212–13
  tensile strength 13, 189
microsilica 46
moisture, damage by 145
mortars
  mixes 136
  plasticisers 136–7

neutrons 23
nickel in steel 191
niobium in steel 191
normalising of steel 192
nylons 287

oxygen 25

paints
  alkyd 300–301
  bituminous 302
  cellulose 302
  driers 300
  emulsion 301
  extenders 300
  functions 298
  microporous 304–5
  oil-based 300–301
  pigments 300
  polyurethane 301
  saponification 301
  types 299–300
  vehicles 300
particle boards 257–9
pearlite 190–93
periodic table 27
permeability 145
phenol formaldehyde 289
phosphor bronze 202–3
plasterboards 179–81
plasters
  defects 181–3
  for damp applications 181
  lightweight 177–9
  types 175–9
plastics (see also individual types)
  in fire 17, 291–2
  painting 305–6
  recycling 294–7
  thermoplastics 280–84
  UV degradation 16
  thermosetting 284
plywood 259
polyamides (nylons) 38, 287
polycarbonate 287
polychloroprene rubber (neoprene) 39,
  289
polyesters 290
polyethylene (polythene) 285
polymerisation 279–80
polymethyl methacrylate 287, 293–4
polypropylene 17, 285
polystyrene 285–6, 292

polytetrafluoroethylene (PTFE)   39, 287
polyurethane   291–4
polyurethane paint   301
polyvinyl acetate (PVA) adhesives
   139
polyvinyl chloride (PVC)   17, 293–4
polyvinyl fluoride   287
pore structure   144
porosity   141–2
Portland cements   42–51
pozzolanic cements   46, 87
preservatives for timber   23, 231–9
protons   22–4
pulverised fuel ash (PFA)   46

quality
   control   6
   management   7
quenching of steel   192–3

radon in dwellings   17
recycling   18–20
reinforcement for concrete   196–9
renderings   139–41
road pavements
   defects   318–19
   design   312–14
   surface dressings   317–19
rubber   13, 288–9

safety
   fire   14–15
   health   2, 5, 17
   structural   2, 5, 8–10
sealants   288–9
silicon   25
silicone rubbers   289
specific surface   44–5
standards, development of   3

steel
   brittleness   193–4
   corrosion   208–10
   definition   190
   ductility   194–5
   fibres   37
   galvanising   210
   heat treatment   191–2
   high-yield structural   196
   hot-rolled structural   13, 196
   impact strength   193–4
   low-carbon   191
   oxidation   13
   prestressing   199–200
   protection   209–11
   reinforcing   196–9
   stainless   202–3, 210
   structural   195–200
   welding   195, 204–5
stone, natural   132–4
strength of materials
   definition   8
   measurement of   10–11
sulphate attack on mortar   127
sulphates in soils   86–7
surface tension   142–4

tars   311–12
thermosetting plastics   289–91
timber
   adhesives for   260
   air drying   226
   conversion   226–7
   creep   260
   distortion   227–8
   durability   16, 228–31
   environmental aspects   260–63
   fire properties   239–42
   fungal attack   228–30

glulam   241, 259–60
hardwoods   222–3
heartwood   223
identification   263–7
insect attack   230–31
moisture in   223–6
moisture movement in   225, 267–8
paints   304–6
preservatives   231–9
products   255–60
recycling   261–3
sapwood   223
seasoning   226
softwoods   220–22
stains   305
strength   13, 242–50
stress grading   243–7, 268–71
structural uses   242–55
structure   219–23
thermal properties   240
varnishes   305
toughness, definition of   11

unsaturated compounds   278–9
urea formaldehyde   17, 146–7

van de Waals bond   32, 280

water
   crystallisation during freezing   125–7
   repellent coatings   151
welding
   PVC   286
   steel   195
wood stains   305–6

zinc
   paints   303
   pure   202–3